ENGINEERING EDEN

JORDAN FISHER SMITH

ENGINEERING EDEN

THE TRUE STORY OF A VIOLENT DEATH, A TRIAL, AND THE FIGHT OVER CONTROLLING NATURE

CROWN
NEW YORK

Published in the United States by Crown, an imprint of the Crown Publishing Group, a division of Penguin Random House LLC, New York.
crownpublishing.com

Crown and colophon is a registered trademark of Penguin Random House LLC.

Library of Congress Cataloging-in-Publication Data
Names: Smith, Jordan Fisher.
Title: Engineering Eden : the true story of a violent death, a trial, and the fight over controlling nature / by Jordan Fisher Smith.
Description: First edition. | New York : Crown Publishing, 2016.
Identifiers: LCCN 2016008169| ISBN 9780307454263 (hardback) | ISBN 9780307454287 (ebook)
Subjects: LCSH: Yellowstone National Park—Management—History—20th century. | Yellowstone National Park—Environmental conditions—History—20th century. | Nature—Effect of human beings on—Yellowstone National Park—History—20th century. | Bear attacks—Yellowstone National Park—History—20th century. | Violent deaths—Yellowstone National Park—History—20th century. | United States. National Park Service—Trials, litigation, etc. | Negligence—United States—History—20th century. | Trials—California—Los Angeles—History—20th century. | National parks and reserves—United States—History. | Environmentalism—United States—History. | BISAC: SCIENCE / History. | NATURE / Ecosystems & Habitats / Wilderness. | NATURE / Environmental Conservation & Protection.
Classification: LCC F722 .S643 2016 | DDC 978.7/52033—dc23 LC record available at http://lccn.loc.gov/2016008169

ISBN 978-0-307-45426-3
eBook ISBN 978-0-307-45428-7

PRINTED IN THE UNITED STATES OF AMERICA

Book design: Anna Thompson
Jacket design: Elena Giavaldi
Jacket photographs: Howard Quigley (man holding bear); Bozeman Daily Chronicle *(burning forest); The Aldo Leopold Archives (A. Starker Leopold)*
Map illustrations: Jeffrey L. Ward

10 9 8 7 6 5 4 3 2 1

First Edition

For James and Emma

CONTENTS

PART III: TAKE IT EASY

PART IV: HUMAN NATURE

PROLOGUE

IN THE SPRING of 1972, the chronic pain in Harry Eugene Walker's right arm had come to coexist with such a yearning for freedom and self-determination that it was hard to distinguish one ache from the other.

Harry was twenty-five, and he had been raised since early boyhood to succeed his father as owner and manager of a family farm in northern Alabama. His labor was critical to the farm's survival, yet the farm didn't make enough for him to have his own place. So he stayed in his childhood room in the little white house on a hill overlooking the lower pasture, where he came to chafe against his mother's criticism and attempts to direct his life. Because money was so short, in addition to working on the farm, Harry had other jobs: among them, as an equipment operator for a construction company and part-time soldier for the National Guard.

The ache in the elbow and the ache for breathing room came to Harry at all times: rolling over in bed at night, pitching a hay bale, reaching under a cow to hook up the milking machine. It hurt when he moved the levers on a backhoe for the construction company and when he saluted his commanding officer at the National Guard, whose authority he had come to resent even more than that of his mother.

The pain got bad toward the end of 1971, and favoring his right arm led to muscular pain in Harry's neck and back. He went to the hospital, and the doctor who injected his elbow with cortisone and gave him a cervical collar said Harry would need to take up more sedentary work. But Harry didn't see how, yet. People depended on him.

It's not uncommon for a rebellion of the body to a way of life to be treated solely as a medical problem, and in the spring of 1972 Harry had surgery on his elbow. But because nothing else had changed, less than a month into what would have been a four- or five-month rehabilitation, he was called into work at another of his jobs, where the weakness in his arm seems to have contributed to causing a minor traffic accident in his employer's vehicle.

Harry had never taken a real vacation. He had been talking with his father about having some time off to think about things. On Tuesday, June 6, 1972, someone offered him a ride out of town, and Harry left Alabama, headed north, with no exact destination in mind.

There is hereby created in the Department of the Interior a service to be called the National Park Service . . . which purpose is to conserve the scenery and the natural and historic objects and the wild life therein and to provide for the enjoyment of the same in such manner and by such means as will leave them unimpaired for the enjoyment of future generations.

—THE NATIONAL PARK SERVICE
ORGANIC ACT OF 1916

Ecosystems are not only more complex than we think, but more complex than we can think.

—FRANK EGLER, PLANT ECOLOGIST

WESTERN UNITED STATES

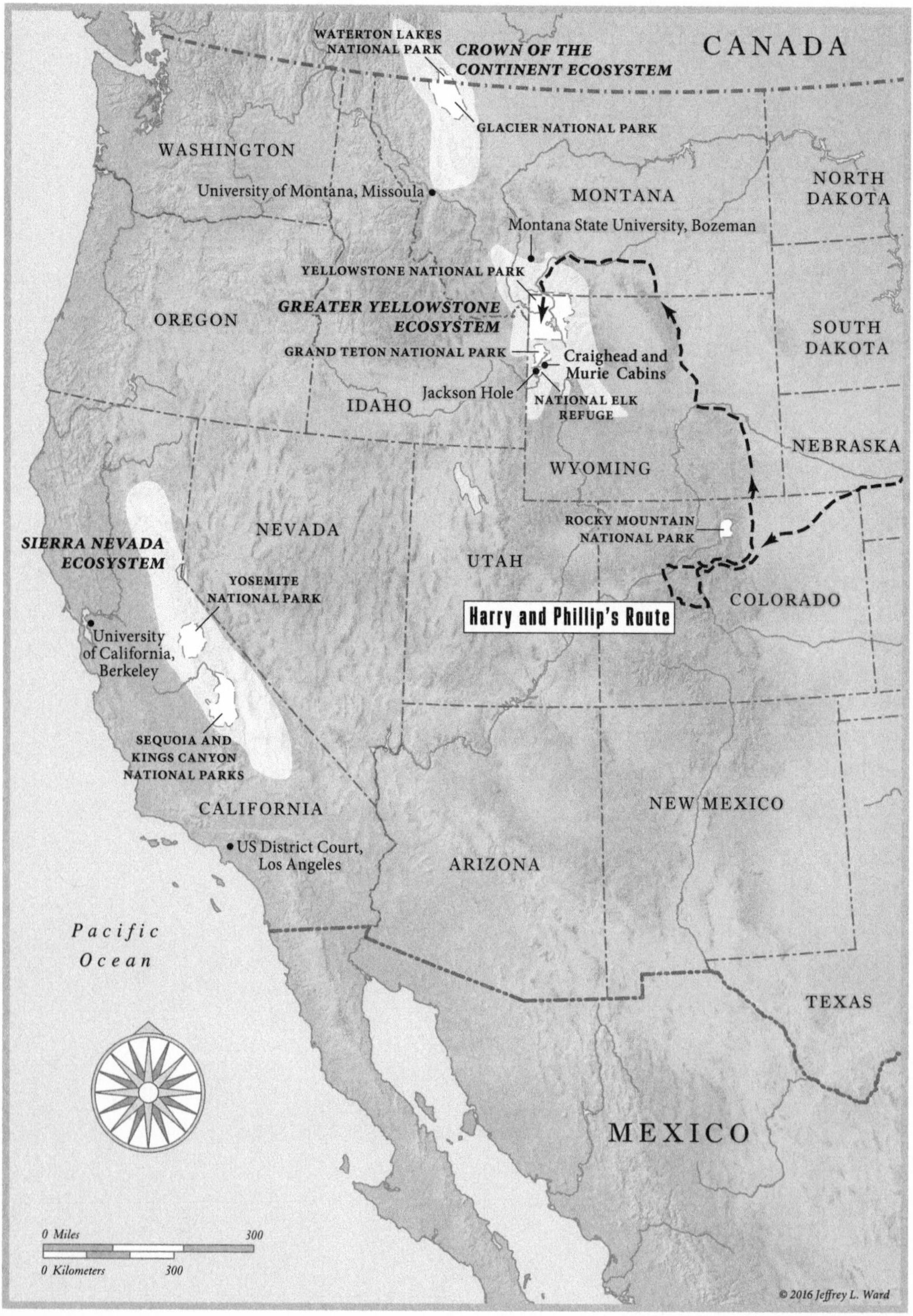

YELLOWSTONE NATIONAL PARK, 1967

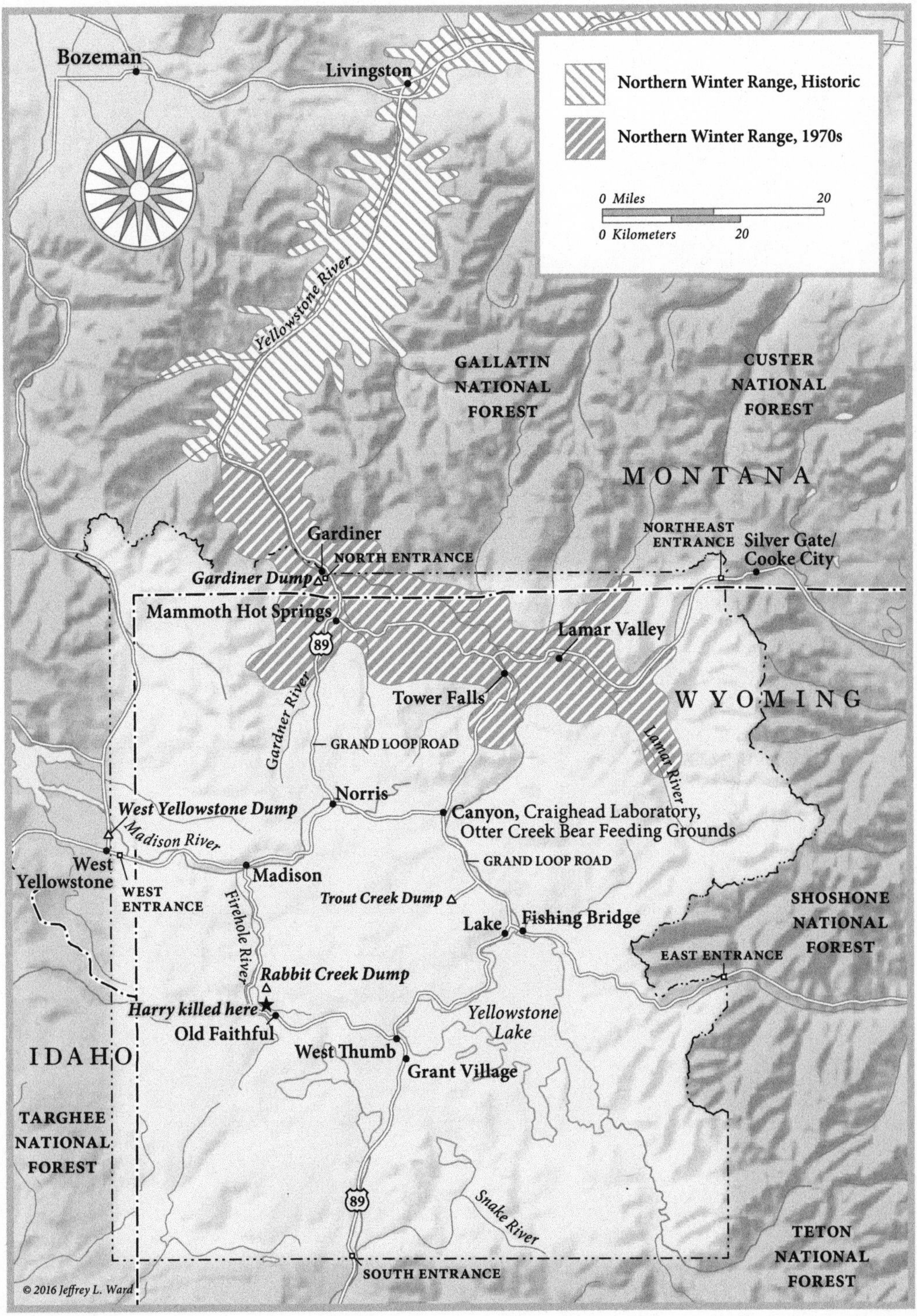

Northern Winter Range, Historic
Northern Winter Range, 1970s
0 Miles 20
0 Kilometers 20
Bozeman
Livingston
Yellowstone River
GALLATIN NATIONAL FOREST
CUSTER NATIONAL FOREST
MONTANA
Gardiner
NORTH ENTRANCE
Gardiner Dump
NORTHEAST ENTRANCE
Silver Gate/ Cooke City
Mammoth Hot Springs
89
Lamar Valley
Tower Falls
WYOMING
Gardner River
GRAND LOOP ROAD
Lamar River
Norris
Canyon, Craighead Laboratory, Otter Creek Bear Feeding Grounds
West Yellowstone Dump
Madison River
West Yellowstone
WEST ENTRANCE
Madison
GRAND LOOP ROAD
Trout Creek Dump
Firehole River
Lake
Fishing Bridge
SHOSHONE NATIONAL FOREST
EAST ENTRANCE
Rabbit Creek Dump
Harry killed here
Old Faithful
Yellowstone Lake
West Thumb
Grant Village
IDAHO
TARGHEE NATIONAL FOREST
89
Snake River
TETON NATIONAL FOREST
SOUTH ENTRANCE
© 2016 Jeffrey L. Ward

PART I

AMERICAN EDEN

1

LOS ANGELES

"ALL RIGHT. CALL the matter," said Judge Andrew Hauk to the court clerk, seated below and in front of him.

"Seventy-two-dash-three-zero-four-four, Dennis G. Martin versus the United States," announced the clerk.

It was a Thursday morning, the ninth of January, 1975, when the trial concerning the death of Harry Walker, known by then as *Martin v. United States,* convened in United States District Court in downtown Los Angeles. The courtroom was an impressively large chamber with fluted mahogany pilasters at intervals along its hardwood-paneled walls, their capitals touching a high ceiling of acoustical tile and fluorescent lights. A small audience was scattered in three blocks of hardwood pews, separated by a low fence from the judge, lawyers, court clerk, court reporter, and bailiff.

A tall lawyer in a fine suit with an unruly head of curly, salt-and-pepper hair stood up from his chair at the leftmost of the two attorney's tables at the front of the room.

"Stephen Zetterberg for the plaintiffs, Your Honor," he said.

"William Spivak, Your Honor," said the assistant US attorney, rising from his seat at the defense table to the right. He was an owlish, balding man in his thirties with glasses. "For the record, I would like to renew my objections to the venue," he added.

Spivak was referring to a highly irregular maneuver by which Zetterberg had gotten a case about an Alabaman who died in Wyoming adjudicated in a Los Angeles court that normally would have had no jurisdiction in the matter. Federal district courts are spotted all over the United States, and a given case will be heard in a particular court when one or more of the parties lives in that district, when the disputed matter took place there, or because assets in the case are located there. None of these things had been true of the Walker case when Stephen Zetterberg took it on, and what he then did was an expression of the creativity he brought to lawyering.

In court, Stephen Zetterberg affected a restrained, dignified manner. Uncoiling his lanky frame to announce his readiness to proceed that morning, he reminded one witness of Abraham Lincoln. But underneath his solemnity he was a passionate man. He had grown up, and still lived and worked, in Claremont, a pleasant university town east of Los Angeles, with tree-lined streets laid out on a gentle slope of alluvium at the base of the San Gabriel Mountains. He graduated from Pomona College and Yale Law School, and during World War II he served on a Coast Guard ship patrolling for submarines out of Pearl Harbor. After the war, Zetterberg returned to the law in greater Los Angeles. By the 1960s his practice was thriving, he was active in politics, and California's governor offered him a series of judgeships. Zetterberg turned them down. He later explained that judges had to take whatever cases came before them, and in private practice he could take the ones that really interested him.

Zetterberg saw the courts as a democratic institution through which the little people could confront powerful adversaries, such as government and corporations. He was attracted to cases involving an underdog. His son, Charles, who became his partner at Zetterberg & Zetterberg after law school, complained affectionately that at any given time his father always had some hopeless matter that could be counted on to bleed the practice of billable hours while the younger associates tried to keep the lights on and make a living. The Walker case was that one in the 1970s.

The case of Harry Walker had three things going for it. First among these were Harry's survivors, the Walkers themselves. To Zetterberg, they were the salt of the earth, *American Gothic* without the dour expressions. Second, they genuinely needed his help. Deprived of their son's labor on their farm, they were in danger of going out of business. Third, there was a great expert witness on their side, a famous biologist who would testify that something had gone terribly wrong with the Park Service's management of nature at Yellowstone. And there was Yellowstone itself. Zetterberg had no personal enmity toward national parks. On the contrary, he loved them. He and his wife were avid hikers, and the fact that the case would involve visits to Yellowstone, Grand Teton, and Yosemite for research and depositions was a major attraction for him. Finally, Zetterberg had already handled two other lawsuits against the Park Service; he knew the case law.

Martin v. United States had acquired its name from Dennis Martin, a young associate lawyer who sat next to Zetterberg at the plaintiff's table. A Yale Law classmate of Hillary Clinton's, Martin had been recruited to the firm in the spring of 1972, in one of Zetterberg's periodic trips back to New Haven to scout promising members of his alma mater's graduating class. Martin was clerking for Zetterberg and hadn't even passed the California bar when he became involved in Zetterberg's scheme to extract the Walker case from its natural venues and bring it to California just so Zetterberg could represent the Walkers.

When the Walker family contacted him about Harry's death, Zetterberg sent Martin to a state court with a motion requesting that Martin be named administrator of Harry's estate. In order for that to happen, at least some of Harry's assets would have to be located in California, however Harry's estate—consisting of little more than a few clothes, a secondhand car, some fishing rods, hunting rifles, a shotgun, and a pool cue—was at his parents' home in Alabama.

Zetterberg's pleadings were a circular arrangement of interdependent *ifs*. Administrators of estates are empowered to take various actions, and Martin told the judge that *if* he were to be so appointed, he planned to sue the federal government for negligence in Harry's death. *If* the estate were to win such a suit, the award would be paid to the estate in California. Therefore, the judgment's potential value could be construed as a California asset—just as an account receivable is listed as an asset on the balance sheet of a business. *If* such an asset could be construed to exist, then the estate had California assets, and it could sue in a California court to create the judgment the whole idea was based on. The argument was a Mobius strip, a snake eating its own tail. Zetterberg referred to the maneuver as "bootstrapping," after the tall-tale notion of reaching down to grab your own bootstraps and lift yourself off the ground.

The state judge apparently admired Zetterberg's fancy and approved Martin as administrator. Zetterberg and Martin then filed suit against the Park Service in Los Angeles. In a preliminary appearance before Judge Hauk, Assistant US Attorney Spivak objected, but Hauk came down on Zetterberg's side. Now Spivak renewed his objection and Judge Hauk defended his decision. There was no—as he put it—"skullduggery" or "callosity toward the law" in what Zetterberg and Martin had done, and he intended to give a fair trial. The case of Harry Walker's death at Yellowstone National Park would be heard in Los Angeles.

"I ruled that way before. I rule that way again," Hauk concluded. "I am going to keep the jurisdiction. I think, therefore, we will proceed."

IT HAD TAKEN over two years for *Martin v. United States* to reach trial, and when it was finally docketed for early January 1975, Stephen Zetterberg's office made arrangements for Harry's father, mother, and youngest sister to travel from their Alabama dairy farm to Los Angeles. It was the first time any of them had ever set foot on an airplane. Harry's mother, Louise, spent the trip to LA in the aisle seat, farthest from the window, gripping the seat arms with a pained expression every time the aircraft lurched over a thermal. Harry's father, Wallace, had grieved no less deeply than his wife over the loss of their only son, but he displayed as much youthful glee at his first view of the earth from above as did his twenty-year-old daughter, Jenny. Hurtling west over Louisiana, the two of them watched, transfixed, out the window as dusk wrapped the earth thirty thousand feet below, even as the aircraft's wings sparkled in the orange sunlight. At Los Angeles International, the Zetterbergs picked them up and installed them in a Claremont resort hotel.

On the first morning in court, Wallace sat listening in the pews. Jenny and her mother were out sightseeing with Stephen Zetterberg's wife. Zetterberg stood facing the judge at the podium between the defense and plaintiff's tables, which the lawyers were required to use when presenting their cases or questioning witnesses.

"All right," said Zetterberg, "next I would like to offer what amounts to—" Here he paused, glancing over his shoulder.

"Would you mind stepping out, Mr. Walker?"

Wallace stood up stiffly and made his way to the exit at the back of the room. Zetterberg waited for the door to swing shut, then finished his sentence.

"—what amounts to an autopsy report of Harry Walker, consisting of seven pages. Mr. Spivak has a copy in his hand."

The document contained the death certificate and pathology report, as well as the typewritten narrative of a strong-stomached Yellowstone National Park wildlife biologist who'd been dispatched to a Livingston, Montana, funeral home to serve as the Park Service's witness to the autopsy. The latter read, in part:

> The body was examined carefully for tooth marks in an effort to measure spacing between canine tooth punctures. Very few puncture wounds were found, however, and none appeared to have been caused by large canine teeth. Most of the injuries seemed to have been caused by claws.
>
> The body cavity and cranium were opened by Dr. Steele; and the brain and body organs not previously removed by the bear were examined. Dr. Steele remarked that other than slight sub-cranial bleeding, which could have been the result of a mild concussion, there was no apparent skull or brain damage. The larynx had been crushed, apparently by a bite to the throat; and

> Dr. Steele felt at the time of the examination that anoxia from this injury, coupled with shock, seemed to have been the cause of death.

"All right," said Judge Hauk from the bench. "The autopsy report of the decedent. Any objections?" he asked Assistant US Attorney Spivak.

"I don't know what this adds," replied Spivak. "It's been stipulated that the decedent died in a bear attack."

Stephen Zetterberg explained that the cause of death, a crushing injury to the neck from the massive power of the grizzly's jaws, yet without the puncture wounds that would normally have been inflicted by the canine teeth, demonstrated that a particular old, toothless bear was involved.

Bears are known for their long memories, part of a general tendency in nature to remember more than it forgets, in layers of stone, in the concentric rings of ancient trees, the migrations of elk, antelope, and trumpeter swans, even in our own recollections of the joys and sorrows of childhood. The autopsy was part of an arrangement of facts with which Zetterberg intended to indict authorities at Yellowstone National Park for believing that nature would forget our past mistakes the minute we tried to remedy them. Zetterberg didn't think nature worked that way, any more than people did. He had watched nature, hiking in the San Gabriel Mountains near his home, and in Yosemite, but he spent much of his working life in court, and courtrooms are full of long-remembered grievances.

2

AMERICAN EDEN

YELLOWSTONE NATIONAL PARK is an approximately rectangular, 2.2-million-acre plot of public land in the northern Rocky Mountains, located in the northwest corner of Wyoming, bordered by Idaho on the southwest, a small strip of Montana to the west, and the bulk of Montana to the north. Founded in 1872, it was the first national park in the world, a pioneering experiment in keeping a beautiful place unaltered against the most fundamental characteristic of human civilization: the alteration of everything we touched.

Located on a volcanic plateau, much of Yellowstone is over seven thousand feet above sea level. The greater part is forested, some in fir and spruce, but mostly in rank upon rank of lodgepole pine. Yet the park is most famous for the smaller portion that is open land: expanses of meadow, sagebrush steppe, and stony ridges dotted with herds of big herbivores—bison, pronghorn antelope, elk, deer, and bighorn sheep—that reflect a mythic sense of what the West looked like before swaths of it were adapted for domestic livestock, alfalfa fields, tree plantations, gas wells, and housing tracts.

The Continental Divide takes an indistinct course across Yellowstone's volcanic highlands, capriciously assigning drainages to the watersheds of the Pacific and Atlantic oceans. The southern part is drained by the Snake River, which flows into the Columbia, and from there into the Pacific. In the north, east, and west, the Yellowstone, Madison, and Gallatin rivers carry the park's waters to the Missouri, and thence down the Mississippi to the Gulf of Mexico. At one curious location near the south boundary, Two-Ocean Creek splits into Atlantic Creek and Pacific Creek. "Here a trout twelve inches in length can cross the Continental Divide in safety," remarked the nineteenth-century fur trapper Osborne Russell.

The name "Yellowstone"—which referred to the Yellowstone River before it was attached to the plateau at the headwaters, or the park—was in circulation among late-eighteenth-century fur trappers in New France as "Roche Jaune." Anglicized as the River Yellow Stone, the name appeared on an 1805 map dispatched to President Thomas Jefferson by Lewis and Clark. By the late nineteenth century it had become associated with the rich ocher color of the rock in the 1,500-foot-deep Grand Canyon of the Yellowstone in the north-center of the park, but it may have referred originally to the pale yellow sandstone bluffs the river passes downstream, in eastern Montana.

Yellowstone is ringed by mountains—the Absarokas, the Beartooths, the Tetons, and the Madison and Gallatin ranges, but its middle gives an overall impression of flatness. That was the impression Lieutenant Gustavus Doane got of it on a summer day in 1870, as a thirty-year-old Army officer heading up a protective detail of soldiers on an expedition composed of territorial officials. On the twenty-ninth of August that year, Doane ascended a peak north of Yellowstone Lake, and looking south from on top, he noticed an oval-shaped void in the Rocky Mountains thirty by forty-five miles across. Doane was an educated man, and he guessed its origins. "The great basin," he wrote, "has been formerly one vast crater of a now extinct volcano."

He was right, except the volcano wasn't extinct.

Few of history's contradictions are as striking as the fact that the first really big natural landscape that human beings set out to preserve in perpetuity happened to be sitting on an active volcano that could be expected, within ten years or ten thousand, to vaporize the whole place. Eighteen or more million years ago, a huge plume of molten rock emplaced itself under the western edge of North America, moving inland as the continent was added to by material scraped off the seafloor in the slow-motion collision between the continental and oceanic plates. As the West Coast grew farther away, every once in a while the molten material would force its way to the surface and erupt in a series of what would have been massive catastrophes had there been anyone with real estate interests or insurance policies to compensate for them. Geologists have traced progressively younger deposits of lava and ash from these explosions in a north-bending crescent from northern Nevada across the Snake River Plain through Idaho to northwestern Wyoming. For a couple of million years that material, an underground mass of semi-molten rock 270 miles high, has been parked under Yellowstone. In that time it has produced three gigantic explosions and a series of smaller eruptions and lava flows. The evidence is that the first big one occurred about 2.1 million years ago. The last one, 640,000 years ago, was a thousand times bigger than the 1980 eruption of Mount St. Helens. Geologists have identified volcanic ash from this expolosion as far away as Canada, Baja California, and Louisiana.

The United States government considers Yellowstone an active volcano and has set up a Yellowstone Volcano Observatory to keep an eye on it. But the frequency of supervolcano explosions occur in such big and inexact numbers that it is not possible to say exactly where we are on Yellowstone's schedule. The magma plume is pushing portions of the park upward—at Hayden Valley, about thirty inches in fifty years. Yellowstone Lake has been found to be tilting, inundating trees at one end. The rocks under the park are riven with faults, and the observatory records one thousand to three thousand earthquakes there each year.

AMONG INDIANS, AND then among trappers and explorers, and later among tourists and scientists, Yellowstone was known for its curiosities and wonders. In a few places its forests contain trees turned to stone. Petrified wood is almost never found standing upright, but at Yellowstone volcanic explosions in the distant past buried standing forests.

Then, over millions of years, the trunks of the trees were mineralized, or "petrified." Later, the soft stone around them eroded away, leaving broken-topped groves of 48 million-year-old extinct sequoia and pine trees with all the fine detail of their bark intact, standing amid living pines and firs.

The most notable effect of the magma chamber beneath Yellowstone is what happens as water seeps down toward it through cracks in the rocks. Yellowstone contains over half the world's geysers, as well as bubbling mud pots, hissing steam vents, terraced travertine fountains, and pools of hot water painted in garish shades of aqua, yellow, red, and orange by cyanobacteria adapted to temperatures that would kill most other organisms on earth. Clouds of steam rise from the forests and meadows, and in the winter the bare ground of thermal areas provides steamy, high-altitude winter range for bison and elk that normally would have to descend to lower, more protected areas. Standing around amid the fumaroles, bison are sheeted with a hundred pounds of rime and icicles. Portions of the Grand Loop Road running past Old Faithful remain snow-free all winter without plowing. Steam explosions hurl rocks and spit gravel. One geyser that erupts only every few years shoots water three hundred feet in the air.

Human civilization has a simplifying effect on ecology. From the tallgrass prairie of the Great Plains, we made wheat fields. From the rich marshes of Florida, we made tomato fields. Located in one of the more remote areas of the West, developed late, and saved early, Yellowstone was not simplified. In 1972, the park had sixty-six species of mammals, all but one of those present in 1850. Today, with the reintroduction of wolves, it has all of them. Two hundred and eighty-five kinds of birds are seen there. In less than sixty-three air miles, north to south, radically different climates support a range of plant communities. Ten inches of annual precipitation fall on some of the sagebrush hills up north. The Bechler River country in the park's southwestern corner gets eighty. The wet places are a riot of color in the spring, with spikes of purple monkshood, crimson paintbrush, pink geraniums, yellow arnica, and white bog orchids standing along misty riverbanks patrolled by fish-eating osprey and sandhill cranes. So are the dry places, with yellow balsamroot and electric blue low larkspur. The bloom is followed by a wave of fruition: blueberries, orange umbels of mountain ash, elderberries, chokecherries, and pine nuts.

TO EARLY EURO-AMERICAN visitors, in comparison to New England, Yellowstone certainly looked like a wilderness. But it had been under some kind of human influence for thousands of years before it became a nature-management kindergarten for an otherwise highly advanced civilization that had by then laid a telegraph cable across the bottom of the Atlantic between Ireland and Nova Scotia. In 1959 an eleven-thousand-year-old spear point was discovered during excavation for a new post office in Gardiner, Montana, on the park's north boundary. About four years later, a ten-thousand-year-old stone projectile point was recovered in southeastern Wyoming, and its minerology traced back to Neolithic toolmakers' quarries at Yellowstone. Along the shoreline of Yellowstone Lake, archeologists excavated extensive hunting camps aged at 9,300 years before the present. One recent chief archeologist at Yellowstone estimated there are 80,000 archeological sites in the park, of which only about 1,800 have been documented.

On stone tools recovered from the Yellowstone Lake sites, highly sensitive DNA technology found traces of the blood of bighorn sheep, elk, rabbits, and other game. Hunting pressure on Yellowstone wildlife was probably heavier before the 1700s, when the cold snap known as the Little Ice Age and epidemics of infectious disease reduced Indian use of the Yellowstone Plateau.

Above the Grand Loop Road south of Mammoth Hot Springs, a once-famous industrial zone known as Obsidian Cliff glints strangely in the sun. Formed by volcanic flows high in the mineral silica, volcanic glass from Obsidian Cliff was prized by native toolmakers for the production of razor-sharp knives, scrapers, and projectile points. Sourced from different deposits, obsidian looks about the same, but depending on where it comes from, its chemical makeup differs. This mineral fingerprint allows archeologists to trace stone implements back to where they were quarried.

In Ohio, over 1,400 airline miles from Yellowstone, hundreds of objects unearthed at a Hopewell culture site were made of Yellowstone obsidian. At another excavation in Indiana, blades made of Yellowstone obsidian were found over 1,200 straight-line miles from the park. By the eighteenth century the tribes that inherited Hopewell territory were decimated by European diseases. The trade routes by which their obsidian made its way from Yellowstone to the Midwest may have been, in the words of one archeologist, "vectors of death," transmitting obsidian east and deadly microbes west, ahead of white explorers. Contagion came in waves, first on foot, and later by steam. A smallpox epidemic spread into the northern plains between 1780 and 1782, and another in 1837, aboard a steamboat traveling up the Missouri River to Fort Union. In all, according to Yellowstone historian Paul Schullery, aboriginal North America suffered at least twenty-eight epidemics of smallpox, twelve of measles, six of influenza, and four each of diphtheria, plague, and typhus.

The first non-Indian we know of to visit Yellowstone was the fur trapper John Colter. On his return from service with the Lewis and Clark expedition, he was recruited by the Missouri Fur Trading Company to survey new sources of animal pelts and pass the word among the Blackfeet about the company's new trading post at Fort Union, later the source of contagion in the 1837 smallpox epidemic. In a remarkable five-hundred-mile solo trek in 1807 and 1808, Colter passed through Yellowstone. After 1826 the area was visited regularly during the fur trade, and according to accounts from that time, the Blackfeet, Crow, Sheepeaters, Bannock, and other Shoshone groups were sharing the area for hunting, fishing, and quarrying obsidian.

After microbes did their work, the founding of the national park took place against a backdrop of military mop-up operations. In 1877, some six hundred Nez Perce men, women, and children passed through Yellowstone, fleeing a massacre by Army cavalry with orders to kill them or force them onto a reservation. In a strange juxtaposition of Yellowstone's past and its ecotourism future, the Nez Perce encountered park visitors on camping excursions whom they took as hostages and, in some cases, shot. The following year the US Army campaigned against the Bannock in the region, and in 1879 against the Sheepeaters in what is now the Frank Church–River of No Return Wilderness, to the west in central Idaho.

When this dark chapter in American history was over, by the twentieth century, visitors from Chicago or Great Falls could stroll up a Yellowstone trail and imagine themselves as the first humans in a wilderness that had never been entirely free of people since the end of the last ice age. Because Euro-Americans didn't witness the effects of Indian hunting until after Indian populations had been reduced by infectious disease, we can only conjecture about how they functioned in concert with cougars, bears, wolves, and coyotes in regulating the number of prey species, such as bighorn sheep, deer, elk, bison, moose, and antelope.

THE LAMAR VALLEY, an elongated basin of wide-open grassland and sage steppes in the northeast corner of Yellowstone, has long been known as one of the two or three best places in the park to observe wild animals. For most of the twentieth century the valley harbored America's largest herd of wintering elk. The two-lane road from park headquarters to the Northeast Entrance, which traverses the base of the hills on the valley's north side, is the only road open through Yellowstone in the winter. Not many years ago, when the elk came down from the high country with the first snows, people would drive out to the Lamar Valley to marvel at the mass of blondish-brown, furry backs shining in the winter light, the forest of antlers, and the sparkly dust of snow as the elk pawed around for something to eat. The northern elk herd, as they were called, were seen as one of the last great wild spectacles of North America, an intimation of how things had once been, before they were altered. Or so people thought at the time.

A short piece southeast along the road through the Lamar Valley from the cluster of log buildings known as the Buffalo Ranch, there is a paved turnout where visitors get out of their cars with their binoculars and spotting scopes to observe herds of bison and pronghorn antelope. From 1989 to 2013, a Park Service educational placard stood facing the road there at waist level. The text was laid out over a large photograph of what you would see on an average summer day from there: grasslands, a row of old cottonwood trees, and wild animals. The text explained that the Lamar Valley supported "... a remnant of the vast wildlife herds that once roamed North America" above which was the placard's title, in large letters: AN AMERICAN EDEN.

And so it seemed to any visitor who didn't know the place's history. To anyone who did, the Lamar Valley bore less resemblance to Eden than to the Civil War battlefields the Park Service takes care of back east. For decades it was probably the most scientifically contested piece of ground in America. The fight there was about how much scientists ought to manipulate and control nature in order to preserve it.

Arguments are rooted in uncertainty. There is little controversy about things we know for certain. In order to understand the disagreement that began at the Lamar Valley and spread to the rest of Yellowstone we must go back to the early nineteenth century, when what was about to happen to the western United States could be compared to the loss of knowledge of the ancient world when the Library of Alexandria burned to the ground in 48 BCE. But in this case, the "library" that was to be burned—and cut down, dug up, shot out, and sold off—was the information that could have been gathered, had there been anyone with today's ecological skills to do it, about what nature was and how it had worked before it was altered.

3

YOSEMITE AND YELLOWSTONE

IN 1812, AS the United States acquired extensive lands from Indians and European nations by expropriation, war, and treaty, Congress created the General Land Office to survey and dispose of this acreage and, in so doing, to encourage settlement of new territories and create revenue to pay off the cost of the second war with Great Britain. Much of America was forested, and agricultural crops required sunlight, so millions of acres of timber were cleared and burned without any attempt to salvage the wood. Where timber itself was the desired product, loggers chewed through the woods at a furious rate in a boom-and-bust pattern that leapfrogged from one virgin forest to the next. In 1850, New York led the United States in timber production; by 1860 that honor had passed to Pennsylvania; and from there it went to Michigan, Minnesota, and Wisconsin, where in 1866 the Land Office's commissioner worried that the forests of the Upper Midwest were "so diminishing as to be a matter of serious concern." In California, Oregon, and Washington, millennia-old rainforests were cut to build San Francisco, Portland, and Seattle. By the 1870 census, the United States had 26,945 lumber mills employing 163,637 men, and no national laws regulating how many trees were taken by them. Logs were snaked out of the forests using wood-burning steam winches and locomotives, and heating and cooking were done by wood fires, so wherever logging went, catastrophic fire followed. At Peshtigo, Wisconsin, in 1871, a series of small blazes coalesced into a 1.2-million-acre firestorm that killed an estimated fifteen hundred people. In response, loggers tried to get more trees out faster, before their own activities burned them up.

Similar scenes of wanton destruction were taking place in mining. In California in 1852, miners began experimenting with fire-hose-like nozzles to wash gold-bearing mountainsides into sluices coated with toxic mercury. The technique led to the development of water cannons that liquified whole landscapes, leaving eroded pits a mile wide that remain barren today. Winter rains sluiced the mine wastes down from the mountains and into the valleys, burying farms and towns. Pennsylvania and West Virginia were torn up for coal. At Butte, Montana, just northwest of Yellowstone, copper ore was smelted by stacking it in a mixture with logs in heaps the size of city blocks and setting fire to them. The piles wreathed the town and its inhabitants in smoke, sulfides, and arsenic, which killed nearly every living thing on the surrounding hills by the 1890s.

The devastation left by clearcutting and mining, along with the other fruits of unmediated capitalism—slavery, child labor, poverty, lethal working conditions, pollution, and cruelty to animals—was met by calls for progressive reform that, in the case of environmental damage, coalesced into a movement for the conservation of nature. Almost as soon as it began, that movement bifurcated. A utilitarian branch focused on conserving timber and other resources for a sustained yield in what became the national forests. A preservationist branch concerned itself with saving outstanding scenery for public use in outdoor recreation and contemplation, which, its adherents believed, had far-reaching benefits for the moral and physical well-being of citizens and the country as a whole. These beliefs were rooted in Romanticism, a cultural movement expressed in art, music, and literature that rejected the cold rationality of the Enlightenment in favor of emotion. Romantics believed that the human spirit would be ennobled by viewing great art, reading great literature, and listening to Beethoven—as well as by communing with nature in the Hudson River Gorge, at Niagara Falls, or in Yosemite Valley.

It's not surprising, then, that the first mention we have about the idea of a national park came from an artist. In 1832, the painter George Catlin, who became famous for his images of Plains Indians in native dress, took a trip up the Missouri River in the new steamer *Yellowstone.* Debarking at Fort Pierre, South Dakota, he watched Sioux Indians slaughtering buffalo to trade for whiskey at the fort. In that scene Catlin foresaw the extinction of both bison and the Native American way of life. In his notes, he suggested that both the buffalo and the Indians could be preserved "in their pristine beauty and wildness" if the government would create around them "a magnificent park."

ALTHOUGH YELLOWSTONE WAS to become the first national park, the idea of such a place was realized eight years earlier in a state park: Yosemite.* The mixture of political progressivism and Romanticism that would characterize the preservationist branch of conservation could be seen in careers of Yosemite's advocates. Thomas Starr King was a Unitarian minister in Massachusetts and an outspoken opponent of slavery. During the Civil War he became an organizer and fund-raiser for the US Sanitary Commission, a precursor of the Red Cross, chartered to care for the war wounded. In 1860 he booked a passage to California to take over a congregation in San Francisco. That July he visited Yosemite, and in the great, glaciated granite gorge he saw the manifest glory of God. On his return to San Francisco he preached a series of sermons about Yosemite and published an influential series of letters about his trip in the *Boston Evening Transcript.*

King's ally in the effort to create a Yosemite reserve was Frederick Law Olmsted, codesigner of Central Park, a fellow abolitionist, and another organizer of the Sanitary Commission. In 1863, like King, he went, to California, where he was employed as administrator of the military adventurer John C. Frémont's Mariposa mines. From there

* A national park was created around the existing state park in 1890, and the state park was eventually annexed by it.

Olmsted came to know Yosemite Valley and the ancient giant sequoias of the Mariposa Grove, and he joined forces with King and with Frémont's wife, Jessie, to lobby Congress to save them from commercial exploitation by having them deeded to the state of California as a public resort.

The Yosemite grant slipped through the national legislature with little comment in the final, bloody year of the Civil War. As President Abraham Lincoln signed the legislation in Washington on June 30, 1864, burial details were attending to some four thousand dead from the Battle of Kennesaw Mountain, in Georgia, three days before. The new law authorized the federal Land Office to grant Yosemite Valley and the Mariposa Grove to the state "with the stipulation . . . that the premises shall be held for public use, resort, and recreation."

The state appointed Frederick Olmsted to make recommendations for managing the new park. In his 1865 preliminary report, Olmsted attested to the power of restful recreation and contemplation of natural grandeur to elevate and refresh the human spirit and heal physical infirmity. This effect was no secret to Europeans, he wrote. In England wealthy elites had acquired private estates in order to reap the benefits for themselves and their handful of guests, but not for the public at large. With the rapid passage of western lands into private ownership, unless the government took deliberate action, America's most inspiring scenery would soon be monopolized by the privileged classes, warned Olmsted. It was fundamental to the democratic ideals of the United States that such places and their benefits be owned and enjoyed in common, he argued.

At the time, only a few hundred people a year visited Yosemite, but Olmsted predicted that a century hence, visitors would number in the millions. He recommended construction of a good road into Yosemite, not only to ease travel for visitors, but also to encourage procurement of lumber and other materials from outside, to prevent damage to the park. He urged that only such facilities as were absolutely necessary be constructed, and that those be designed and sited in such a way as to be as unobtrusive as possible. Laws and regulations should be developed and enforced to keep the park for posterity "in its present condition," as "a museum of natural science."

IN THE MONTANA Territory, by the time of the Yosemite land grant, precious-metal strikes throughout the West had brought prospectors to the headwaters of the Yellowstone River, and they followed the trappers in exploring and mapping the Yellowstone Plateau. The miners were in turn followed by a series of government and quasi-governmental expeditions for mapping, exploration, and scientific purposes.

The first of these was undertaken by a delegation of territorial officials in August and September of 1870. The nineteen-member expedition included Henry D. Washburn, a former major general in the Union Army and surveyor general of the Montana Territory; a Montana lawyer by the name of Cornelius Hedges; Truman C. Everts, a federal tax assessor; a businessman, bank examiner, federal tax collector, and would-be Montana gubernatorial candidate, Nathaniel P. Langford; and a protective detail of Army cavalry led by Lieutenant Gustavus P. Doane, who during this expedition would make his observation of the great gap in the Rocky Mountains that is today known as the Yellowstone Caldera.

For much of the twentieth century the National Park Service clung to the account of expedition member Nathaniel Langford that the idea of the national park did not spring from the minds of George Catlin, Frederick Olmsted, or Thomas Starr King, but was suggested during a conversation on the evening of September 19, 1870, among members of this expedition. According to Langford, they were talking around a campfire in the valley where the Gibbons and Firehole rivers flow into the Madison River, when some of them began discussing opportunities for commercial development of Yellowstone. Langford contended that Cornelius Hedges spoke up against this, suggesting that the government should secure the area as a public reserve. In the course of the discussion the others came to agree with him.

The Park Service preferred this story because it featured the altruistic invention of the national park, within the first national park itself. But there is no mention of this conversation in the journal entries of the men involved, and Langford gave no account of it until twenty-four years later, and then in only the most cursory manner. The full tale did not appear until Langford's publication of a re-created "diary" of the expedition in 1905, thirty-five years after it supposedly happened.

The actual motive behind Yellowstone's charter appears to have been more pecuniary. The financier Jay Cooke, one of the richest and most powerful Americans of the nineteenth century, had underwritten the United States's $2.6 billion debt to fund the Civil War. Cooke's relations with the government were sufficiently cozy that the construction of one of his interests, the Northern Pacific Railroad, was incentivized by the most generous government land grant ever bestowed on an American railroad. In June of 1870, just before the Yellowstone expedition, Langford visited Cooke at Cooke's private estate. There is no record of what they talked about, but the prospect of a spur line to a new park full of wonders would surely have created excitement around the sale of the railroad's bonds. Langford's subsequent activities bear out the nature of their relationship.

After the meeting Langford spearheaded the expedition to Yellowstone. On his return, Cooke paid for him to go on a speaking tour, and Langford authored articles for *Scribner's Monthly* extolling Yellowstone's wonders. To illustrate them, *Scribner's* commissioned the English Hudson River School painter Thomas Moran, who, having never visited the area, had to base his work on Langford's notes and sketches. This evidently made Moran eager to see the place himself.

In the audience at Langford's first lecture in Washington, DC, was Ferdinand V. Hayden, a doctor, self-trained geologist, and explorer who had been conducting mapping and scientific expeditions for the US Geological Survey, and had previously tried, and failed, to get into Yellowstone. Spurred by Langford's talk, in 1871 Hayden secured government funding to mount another expedition. He invited the pioneering western photographer William Henry Jackson to go along. As Hayden started out for Yellowstone from Utah in the spring of 1871, he received a request from Cooke's office manager asking him to take along Moran, as well. They rendezvoused along the way, in Montana.

The credulity of the American public about the bizarre and beautiful features of Yellowstone had been tainted by the "tall tales" tradition of the fur trappers, who amused themselves with hyperbolic stories, of which Osborne Russell's about a trout swimming across the Continental Divide was one of the truer examples. Jackson's photographs would prove that what were—to people in gentle New England—nearly unbelievable accounts of Yellowstone's steaming grandeur were true. For Moran's part, if Jackson's work was limited to black and white, he had color, and he used it gloriously. Moran exercised seemingly infinite patience in rendering the visual complexity of nature. His tiny brushstrokes illuminated every cleft in the rock, every scab of pine bark, each whirlpool in rushing turquoise waters. Employing successive layers of sunlight and shadow, his landscapes recede into clear-aired infinity. Nevertheless, making watercolor sketches at the Grand Canyon of the Yellowstone, he despaired at trying to capture the golds, ochers, and reds of the canyon walls. He underestimated himself, and a huge oil painting of the canyon's Lower Falls executed after he returned home was purchased by the government to hang in the national Capitol as a symbol of the grandeur of the American land.

On his return to Washington, Hayden received a letter on Jay Cooke's stationery from an agent for Cooke and the railroad suggesting that Congress designate a public park around the geysers of Yellowstone. Hayden, Langford, and a representative from Montana Territory lobbied for the park that winter, distributing copies of Langford's *Scribner's* article to senators and congressmen. Moran's watercolor sketches and Jackson's photographs, which Hayden arranged to be exhibited at the Capitol, have been widely credited with selling the idea. The bill creating Yellowstone "as a public park or pleasuring-ground" cleared Congress and was signed by President Ulysses S. Grant on March 1, 1872. The new law ordained "the preservation, from injury or spoliation, of all timber, mineral deposits, natural curiosities, or wonders within said park, and their retention in their natural condition." This was the beginning of the long and repeated use of the word *natural* in defining the goals of national parks and wilderness areas and the means of their achievement. For a long time, the definition of the word seemed self-evident. But by the time Harry Walker left his family for Yellowstone, what *natural* meant, and how naturalness would be accomplished, was becoming increasingly hard to pin down.

4

APPALACHIAN SPRING

WHERE HARRY EUGENE WALKER grew up, at the southern tip of the Appalachian Mountains, the range bends southwest to terminate in northern Alabama in a pattern of parallel, pine-covered ridges separated by green agricultural valleys. The Walker farm was in one of these drainages, the Choccolocco Valley, about a twenty-minute drive outside the redbrick factory town of Anniston. Choccolocco—local people omit the word *valley*—was a patchwork of cotton fields, cow pastures, groves of loblolly pine on the high ground, and hardwood thickets in the bottomlands. Choccolocco Creek ran down its center, spilling lazily over broad, stair-stepped bedrock ledges that southerners call shoals.

Harry's father, Wallace Walker, was the son of a Choccolocco cotton farmer. By all accounts, since his boyhood, all he ever wanted to do was farm. However, by the time Wallace graduated from high school, repeated cultivation of cotton had exhausted some of Alabama's soils, so rather than following his father into cotton, Wallace apprenticed on the dairy farm of an uncle in Huntsville. There he met and married Harry's mother, Jessie Louise Campbell.

Wallace was a gentle man with a sweet personality. Jessie, known to everyone by her middle name, Louise, was a firecracker. She was tiny—five foot two and 110 pounds—but she had a large, and sometimes very sharp, presence. Nevertheless it was a good marriage, and she and Wallace would be devoted to each other for life. In 1945 Wallace took Louise home to Choccolocco, where they settled in a white, single-story, hip-roofed, three-bedroom farmhouse on a knoll overlooking 110 acres of hardwoods Wallace was clearing for pasture along Choccolocco Creek. He acquired his first twenty-six cows, a used milk-delivery truck, and equipment to set up a creamery, and began delivering milk in refillable glass bottles to residential customers in Anniston and Oxford, the neighboring town to the south. Louise gave birth to three daughters and a son. Harry was the third, born in July of 1947.

As his parents' only male heir, Harry was expected to someday take over the farm. The certainty of his destiny was reinforced by everyone around him from the time he was tiny. The steps outside the back door of the Walker's farmhouse afforded a panoramic view of the pasture Wallace had cleared between the low hill on which the house stood and Choccolocco Creek. The pasture was as flat as a sheet of glass, bright green, and dotted with Wallace's cows. When he wasn't otherwise engaged, Wallace had a habit of standing on the steps with a pleasant, faraway expression, admiring his farm. At the age of three, his son began copying him. Each morning Harry would climb out of bed and toddle purposefully past his mother and older sisters as they prepared breakfast and out the kitchen door to the back steps, where he would conduct his own solemn inspection of the farm as the sun rose over the hay barn, the silo, and the lower pasture. Louise Walker took great pleasure in this. As she saw it, she was cultivating the dairy's future in her little boy as surely as Wallace was in his pasture and herd.

Wallace's own father had begun teaching him to farm when he was eight, and when Harry turned eight, Wallace began teaching him. Patiently, he instructed Harry in each of the dairy's jobs: caring for the herd, growing and harvesting feed corn, cutting and baling hay, and operating the harrows, seed drills, brush hog, feed mill, hay binder, wagon, and other implements hooked to a red Farmall tractor and, later, to a green John Deere. Harry attended 4-H and Future Farmers of America, and by the time he suddenly left the valley in 1972, he had lived most of his life by the heartbeat of the dairyman's days: morning and afternoon milking.

At three o'clock each morning, Wallace would get up to bring in the cows. The bedroom window of Harry's youngest sister, Jenny, faced the fence where the Holsteins gathered during the night. Jenny would wake halfway to the sounds of her father whistling to his dogs and the puffing and blowing of those cows that were lying down as Wallace prodded them to their feet. When he had gotten the animals moving and turned on the pumps and the radio in the milking barn, he'd wake Harry by calling in his sweet, low voice from the front porch outside Harry's bedroom window, around the corner of the house from Jenny's, in exactly the same way each day:

"Hey, Harry."

Harry would pull on his clothes and rubber boots and shuffle across the darkened gravel driveway, into the bare-lightbulb glare of the milking barn, a long, low, concrete-block building painted white, with a row of casement windows facing the house. It was noisy inside with the lowing of the cows waiting at the gate at one end, the rattle and sigh of the compressor for the refrigeration equipment in the tank room at the other, and a portable radio playing rock and roll and country.

When everything was ready, Harry and Wallace would open the gate and the first four cows would amble in, turn right, and put their heads through a row of stanchions to munch on grain from a trough. Harry and his father would sponge off the udders and slip the chrome cups at the ends of the milking machine's spiders of black rubber tubes over the pink teats, and the tubes would jerk rhythmically as the milk dribbled in. As each cow was finished, the man and the boy would bring in the next, stumping through the slippery manure on the concrete floor and singing along with the radio. When the first light shone through the dusty windows, Harry would walk back to the house to get ready for school while Wallace cleaned the equipment and hosed down the floor.

On summer days when Harry wasn't in school, the Walker men would come in from milking, eat breakfast, and disappear into the fields. The farm produced most of its own feed in the form of hay, corn, and silage. There were fences and machinery to fix. Cows got sick; calves got scours. Wallace Walker was known as an accomplished cow doctor. The milk tank had to be scrubbed out—Harry's sisters sometimes did this. At any given time, just under a quarter of Wallace's herd might be out of production, and those cows were "freshened" by breeding them. When they dropped their calves, Harry and his father might stay up all night delivering the slick little beasts. Sometimes they found themselves in arm-length rubber gloves, bicep-deep in a cow's uterus, helping to sort things out when an umbilical cord got wrapped around a calf or some other thing went wrong. The family kept chickens for eggs. Wallace had caught a few Asian koi that someone else had introduced into a nearby creek, and he put them in his watering trough, where they kept the water clear for the cows by consuming algae and insect larvae. Barn cats kept the mice and rats under control, and Border collies and heelers waited at the kitchen door for Wallace to give them something to do. They were so well trained that he could tell them to go down and cut a certain cow from the herd by name—they all had names—and bring it back to him to be doctored. One of the dogs sat expectantly across from the stalls at milking until Harry or Wallace sent a stream from one of the teats into its mouth.

A farm is a manipulated ecosystem, complex in its own right yet far simpler than the famously diverse Appalachian woods Wallace had cleared to make way for his dairy. From the grasses that grew in the lower pasture, to the cows that came up from it at night to wait for Wallace to milk them, to the dogs that herded the cows, the koi that cleaned the watering trough, the plants the Walkers harvested for silage, the farm machinery that harvested the plants, and the bacteria that fermented the silage to release its nutrients, the Walker farm was a ballet of men, women, plants, animals, and implements, each element seeming to know exactly what was expected of it. Farming wasn't always easy, successful, or environmentally sustainable, but a well-run family farm such as the Walkers' was the result of nine millennia of human experiments with domesticated animals and cleared land. A lot more thought had gone into how human beings lived with cows, horses, dogs, feed corn, and alfalfa than how they could live with wild animals and untamed land.

IN 1859, CHARLES Darwin observed that plants and animals were "bound together by a web of complex relations." As an example, Darwin related an account he had heard of the effect of house cats on the seed production of clover. For clover to set seed required the services of a ground-nesting bee, he explained, and field mice tended to eat the bees' nests. When the population of field mice got too high, the production of clover, a protein-rich feed for cattle, would cease. However, close to villages, where housecats were present—Britain's wild predators were by then much reduced—the fields produced a healthy crop of clover. John Muir put this notion succinctly a few years later: "When we try to pick out anything by itself, we find it hitched to everything else in the Universe."

Unfortunately, the study of these complex relations would not emerge as a distinct field until about the same time as atomic physics did, in the 1890s—very, very late in history. Meanwhile, the process of obscuring whatever had been present in the oldest national parks was well under way before they could be studied. In fact, it was going on before the parks were even created, with the genocide of Native Americans, whose observations might have added immeasurably to the ecological backstory, and with a wave of unrestricted shooting of wild animals Yellowstone historian Paul Schullery has called an "ecological holocaust."

In the late nineteenth century, in addition to their appetite for virgin timber, Americans displayed a Paleolithic hunger for wild meat and animal products, and for the sport of pursuing and killing living things. "Game birds snared and shot during the breeding season; cock-shooting in early July; killing grouse in August . . . shooting game of all kinds . . . slaughter of large game of both sexes and all kinds at every season for their hides, or simply for the sake of killing them, . . ." complained an Army officer assigned to the western frontier in a letter to an eastern magazine.

But the killing was not just for sport. Production of North American wild animal products was a major industry aligned with colonialism and imperial conquest, as was true of African ivory, guano from Chile, and wild fish in the world's oceans. Between the 1850s and '70s the

Hudson's Bay Company auctioned off some 3 million wild beaver pelts, the fur from which was used in the production of felt hats. Beginning in the 1870s, fashionable hats for women were decorated with feathers and sometimes with whole stuffed wild birds. The ornithologist Frank Chapman was working at the American Museum of Natural History in 1886 when he made a survey of birds in Manhattan, but the avian life he saw was all dead. Chapman walked twice from his office to the fashion district on Fourteenth Street, surveying and identifying the bright-colored plumage, wings, and stuffed bodies adorning the hats of the women he passed. In all, he identified the remains of 174 members of forty different species, including blue jays and bluebirds, woodpeckers, Baltimore orioles, owls, egrets, and lots of terns.

Fresh wild game was available from many butcher shops, and the commercial taking of ducks, geese, doves, passenger pigeons, quail, grouse, venison, elk, antelope, bison tongue, and strings of trout, salmon, and steelhead was unregulated by law. With the use of ice cut from ponds in winter and rapid transportation by rail, the market broadened. Two months before the passage of the Yellowstone bill, the menu for a special Christmas dinner at a Nashville, Tennessee, hotel included roast loin of Montana buffalo in Madeira sauce; haunch of Arizona elk and venison with red currant jelly; Nebraska antelope in sauce Robert; roast leg of Rocky Mountain black bear; canvasback, redhead, and mallard ducks and wild turkey served with jelly; wood duck, blue-winged teal, and prairie grouse with cranberry sauce; and broiled quail on toast.

In the summer of 1872 a Kansas newspaper reported that a thousand professional buffalo hunters were roaming the western plains. From Fort Benton, a main point of shipment, by steamboat, for buffalo hides from the Montana Territory, Lieutenant Doane reported that 135,000 bison had been taken by hunters that year. Out of an estimated 26 million that had once inhabited the western United States, by 1889 25 bison were found in west Texas, 20 in Colorado, 36 scattered across eastern Montana, and about 200—the largest remaining herd south of Canada—had taken shelter in the relative remoteness of the Yellowstone Plateau.

The same heedlessness characterized the market hunting of elk, deer, bighorn sheep, and antelope for their hides, meat, and, in the case of the elk, for their teeth, prized as talismans by members of the Benevolent and Protective Order of Elks. Nine months after the passage of the Yellowstone Act, a witness reported seeing a fast freight train headed down the new Northern Pacific line toward Bozeman loaded with wild animal carcasses from the area of the park.

Thus, with railway transportation of wild animal products to markets and a new park created without practical mechanisms to protect it, began an orgy killing that unraveled Darwin's "web of complex relations," and set into motion cascading consequences that would prove difficult to see, and more difficult to fix.

THE 1872 YELLOWSTONE legislation placed the park under the care of the US Secretary of the Interior, and authorized the secretary to make such rules as were necessary to protect it. But the law itself didn't promulgate those rules, nor did it prescribe a time limit for their creation. Even if such rules had existed, the law did not provide appropriations for a staff to enforce them. Nor did it specify criminal penalties for violations. These failures sound rudimentary now, but no one had ever run a national park before, and it took a while to learn how.

The quality of Yellowstone's early superintendents ranged from wholly ineffectual, in the case of Nathaniel Langford, the first; to rather good, in the case of the second, Philetus Norris. But they all were hampered by a lack of funds, personnel, and scientific knowledge that didn't exist yet. Langford lasted five years, during which he was physically present in the park only three times. George Bird Grinnell, who went on to found the Audubon Society to protect birds from plume hunters in the millinery trade, visited Yellowstone while it was under Langford's care and estimated that three thousand bison and mule deer had been killed there in the winter of 1874–75. In 1875, Langford himself reported that hunters had shot four thousand elk, along with scores of deer, moose, antelope, bighorn sheep, and bison. His successor, Norris, estimated the number of animals shot that winter at closer to seven thousand. An Army general who visited during the same period estimated that four thousand elk had been killed around Mammoth Hot Springs alone. Olmsted's idea for a "museum for natural science" was off to a rocky start.

Norris was far more effective than Langford had been. He constructed roads and trails and a small fort at Mammoth Hot Springs, and he kept careful records of biological and physical phenomena. What he lacked in ecological insight was only typical of his time. In 1877 he was lacing carcasses of winter-killed elk with strychnine to kill wolves and coyotes. Three years later he reported that he had allowed trappers to take hundreds, if not thousands, of beaver to keep them from making travel difficult through low-lying areas flooded by their dams. Norris was eventually driven out of the job by political intrigue. The next three superintendencies were ephemeral, passing in and out of office in the same number of years.

Early superintendents appointed a handful of civilian assistants. Best known of them was Harry Yount, who is sometimes referred to as the first national park ranger. A former guide for Ferdinand Hayden and Yellowstone's first full-time gamekeeper, Yount was a capable man who nevertheless quit in 1881 after barely a year of service. His resignation letter suggested that no single man could properly protect Yellowstone; what was needed was a small, well-organized police force. Prodded by conservationists such as George Grinnell, by then the influential editor of the sportsmen's magazine *Forest and Stream,* in 1883 the Secretary of the Interior issued a directive declaring Yellowstone's animals and natural features protected by law. However, there were no criminal penalties for violating it and practically no one to enforce it, so the depredations continued. In 1886, the secretary finally admitted to the failure of Yellowstone's civilian administration by formally requesting that the US Army impose order at the park.

Even after the arrival of troops that year, park protection was hampered by the lack of criminal statutes. In this absence, the Army carried out extralegal hazing and detention of poachers, confiscating their equipment and sometimes imprisoning them in the stockade. In 1894, Grinnell's *Forest and Stream* published an eyewitness account of the apprehension of a notorious poacher who'd just killed four buffalo in the Lamar Valley. Public outrage finally prodded Congress into enacting a law—the Yellowstone Game Protection Act—making the taking of park animals punishable by fines and other penalties, and the Army set out to enforce it.

Meanwhile, Army administrators began organized efforts to restore game populations by feeding elk, bison, deer, and pronghorn through the winter on hay grown in the park. Elk rebounded rapidly, but in the long struggle with poachers, the Yellowstone bison population dwindled to twenty-five individuals by 1901. Faced with the possibility of the iconic animal's extinction in the United States, President Theodore Roosevelt requested money from Congress to purchase additional bison from Canada and a private herd in Texas. In an early prototype of the sort of interventions called for in today's endangered species recovery plans, from 1907 on, bison were fed and bred in captivity at the Buffalo Ranch, a cluster of log buildings and unusually heavy fences and stock chutes in the Lamar Valley, parts of which can still be seen today. By 1910 a sturdy military base, Fort Yellowstone, had been constructed at Mammoth Hot Springs, and 324 soldiers were billeted throughout the park, effectively eliminating large-scale poaching of wildlife.

THE 1872 ACT creating Yellowstone National Park contained a famously paradoxical mandate, which would be carried forward in a later law creating the National Park Service to take over the parks from the Army. On the one hand, it ordained that Yellowstone was to be a "public park or pleasuring-ground for the benefit and enjoyment of the people." On the other, its minders were instructed to "provide for the preservation, from injury or spoliation, of all . . . natural curiosities, or wonders within said park, and their retention in their natural condition."

These juxtaposed missions of national parks as public playgrounds and nature reserves have become so commonplace that we rarely stop to consider how much in tension with each other they are. We take for granted that millions of people are allowed to wander more or less freely in national parks, where they intermingle with grizzly bears, mountain lions, wolves, poisonous snakes, spiders, alligators, and crocodiles. At a zoo, a much simpler and older arrangement—Nimrud, the capital of Assyria in what is today Iraq, had one in the ninth century BCE—we wouldn't think of allowing those species to be in such unrestricted proximity to people. But the national park idea is some twenty-eight centuries fresher, and the details of this perilous arrangement are still being worked out—at times, in courtrooms like Judge Hauk's.

Regardless, come visitors did, thousands of them by the 1880s. Men in coats, vests, shirts, and ties and women in stiff-bodiced Victorian dresses and stiflingly long, dark-colored skirts over petticoats and dainty boots wandered over the once-lonely steaming travertine terraces at Mammoth Hot Springs, chipping off bits of stone as souvenirs. They gaped at the sulfurous geysers and peered over the dizzying brink into the Grand Canyon of the Yellowstone. They watched the sun set over the pines, and hiked to the top of Mount Washburn, where Lieutenant Doane had made his observation of the great caldera. They looked out upon the trackless loneliness beyond the railroads and breathed deep the aromatic breeze off the western sage. And they wanted to see animals: deer, elk, pronghorn, bighorn sheep, bison, and bears.

In order to provide for all these visitors (and of course for the railroads and hoteliers) the Yellowstone act authorized the Secretary of the Interior to grant leases to private parties for the construction of tourist facilities, and although protective regulations and their enforcement lagged, for-profit development didn't. In 1880 a rough hostel was thrown up at Fountain Flats, north of Old Faithful. It was superseded in 1891 by the 350-room Fountain Hotel, featuring electric lights, steam heat, and hot water piped in from the geothermal pools. A massive Greek revival resort hotel was constructed at Yellowstone Lake the same year. A humble three-story hostel at Mammoth Hot Springs was dwarfed in 1885 by the 250-room, four-hundred-foot-long National Hotel. A hotel went in along the rim of the Grand Canyon of the Yellowstone in 1886. The 140-room Old Faithful Inn was built in 1903 and 1904 and enlarged with a 100-room addition nine years later. Another 150-room wing was added in 1928. Even before these, the kitchen could feed 347 guests. Not to be outdone, the owners of the National Hotel at Mammoth Hot Springs boasted that its huge kitchen range could accommodate fifteen chefs at once.

For all these modern accommodations, the state of the art in waste disposal was about the same as when the 11,000-year-old spear point found along the park's north boundary had been made. At Yellowstone and nearly everywhere else at the time, that which was not useful anymore was thrown in a pile as close as possible to where it was produced. In the hotel kitchens of Yellowstone, whatever was trimmed off a piece of meat or a head of cabbage, anything scraped from the bottom of a pot, everything that came back uneaten from the dining room, and all the cans, bottles, and crates that food came in went into open dumps just out of sight of the hotels.

By 1889, bears—black bears first, then grizzlies—were emerging from the surrounding forests to feed on vegetable and meat trimmings, lumps of leftover gravy, lard, and bread, the rinds of cheeses, and half-eaten éclairs. By 1891, they were making forays into campgrounds, looking for food. By 1910, stagecoach-loads of visitors were encountering black bears waiting along roadsides to be tossed bits of the visitors' lunches. With the end of World War I in 1918, the number of Yellowstone visitors tripled in a year, and many now arrived by automobile. A black bear named Jesse James became notorious for blocking traffic near the West Thumb of Yellowstone Lake until the occupants of passing cars gave it food. Feeding bears from automobiles became an expected part of the national park experience.

It didn't take long for hotel management to turn the bears' interest in garbage into entertainment. At Fountain Hotel, visitors walked about a hundred yards from their accommodations to watch wagonloads of refuse be delivered to the local bears. Bleachers and benches were put up for spectators to watch scheduled deliveries of garbage at "bear feeding grounds" at the Lake dump, the Otter Creek dump near Canyon Village, and Old Faithful. These "bear shows" proliferated in other national parks such as Sequoia and Yosemite. By 1931, an amphitheater constructed at the Otter Creek feeding grounds seated 250. There and at Old Faithful, food was placed on a raised platform, or "lunch counter," like a stage, so everyone could see the bears. As many as five or six hundred cars arrived at the parking area at Otter Creek for the events, and when the seating was exceeded, audiences gathered in standing room.

Giving a wild animal food is perhaps the most potent way to influence its behavior—far more powerful, wildlife biologists would later learn, than stopping the feedings or even causing pain or discomfort in an attempt to reverse that behavior. At Yellowstone and other national parks, the behavior rangers inadvertently taught to bears was to overcome their reticence to approach people and the areas people frequented. How quickly and to what degree that happened varied according to the species of bear.

Two species of bears inhabit Yellowstone. The larger, known in Alaska and Canada as the brown bear and in the Lower Forty-eight as the grizzly, for the grizzled-looking silver-tipped guard hairs on its brown coat, typically weighs between 300 and 500 pounds, with big males sometimes reaching over 700. Grizzlies can be identified by their coat, distinctive shoulder hump, short noses, and flat, almost dished faces, with eyes facing forward. Adult grizzlies' claws are relatively straight.

The claws of the other species, the black bear, are curved. Black bears are not always black; their coats range from black to brown to even blond, and they sometimes have a distinctively shaped white blaze on their chests. They are smaller than full-grown grizzlies, with adult males averaging around 250 pounds and adult females around 150, although when fed on garbage males have been recorded at over 600 pounds. They have longer noses, and the overall shape of their heads are more like a dog's than a grizzly's.

Wild black bears will flee at the approach of a human being, but after repeated food rewards they will approach with growing boldness to steal from a picnic table or be fed by hand. For the first seven decades of the twentieth century, black bears waited along national park roadsides and appeared in camps and picnic grounds for handouts. Tourists loved this, and an early-twentieth-century Yellowstone superintendent who became director of the Park Service, Horace M. Albright, thought of "tame" bears and the bear shows as a desirable attraction at the national parks. However, because of the close proximity of hundreds of large carnivores to millions of people, it was inevitable that there would be problems. As bears lost their fear of human beings, they began tearing up camps and breaking into buildings and automobiles to get food. In 1932, there were 451 such incidents in Yellowstone, the great majority of which involved black bears. Although grizzlies tolerated spectators while feeding at dumps and bear shows, they avoided human beings at other times and almost never approached them to be fed.

Injuries inflicted by black bears were rarely deliberate or predatory. They occurred when bears were startled or grew impatient while being fed, or when they were teased, such as by someone holding food higher than the bear could reach. In the latter case black bears did the logical thing: they used their curved claws to climb up a visitor or they knocked him or her down. Common injuries were slashes and bites to arms, and mangled extremities that had been holding the food. Between 1931 and 1939, there were 527 injuries of this sort at Yellowstone.

Black bears in the forty-eight contiguous United States almost never kill people. Grizzlies rarely do, and when they do, they sometimes seize their victims, drag them some distance from where they encountered them, and feed on their flesh. The turn-of-the-century disappearance of one guest at the Fountain Hotel may or may not have involved a bear. On the evening of July 30, 1900, a bank cashier from Ohio had dinner at the hotel, bought a cigar in the lobby, walked outside to smoke it, and vanished. A search was conducted but no part of him was ever found. There were a small number of unconfirmed accounts of fatal maulings at Yellowstone, and at least one well-confirmed case outside the park. The first confirmed fatal bear attack inside Yellowstone occurred in 1916, when a grizzly that had already injured two other people entered a camp occupied by three teamsters at night. The bear attacked one man and began to eat him alive. His coworkers succeeded in scaring the bear off, and the man was taken by a passing automobile to Fort Yellowstone, where he died. In revenge, the dead man's companions placed a baited charge of dynamite and blew the bear up.

But in general, in the early decades of Yellowstone's history, injuries to people remained the province of black bears and were nonfatal. Grizzlies were content to feed on their natural foods, at the dumps, and at the bear shows, and they otherwise remained reticent to approach human beings. When they did, they were sometimes shot. There wouldn't be another fatal grizzly attack at Yellowstone for twenty-six years, and even then it was another four years before a new law authorized injured citizens or their estates to sue the National Park Service. So it took a long time for the way bears were managed to come before the federal courts, and, ultimately, to be taken out of the exclusive control of the Park Service.

5

FRANK

THE DOORS TO Judge Andrew Hauk's Los Angeles courtroom were padded in red leather, quilted with upholstery tacks, to limit the transmission of sound from the courthouse's second-floor hallway. The cavernous, fluorescent-lit chamber inside them had not a single window, skylight, or any other source of natural light. You would not have known if it was morning or night but for the attorneys' wristwatches and a clock on the back wall. A pair of air-conditioning vents above and to one side of the clock were the only source of fresh air. In that room, on the third day of the trial, Stephen Zetterberg summoned the lifelong outdoorsman he planned to present as his star witness.

"I would like to call Dr. Frank C. Craighead."

The witness rose from his seat in the pews and made his way up the aisle. He walked with a compact, athletic grace. He was only five foot seven but broad in the shoulders, narrow at the waist, his muscles visible through his clothes. At his present age of fifty-eight, the wind and sun had engraved into his otherwise smooth, tanned face a pair of parentheses on either side of his mouth and a stack of equals signs on his forehead. His gray hair was neatly trimmed and slicked back with tonic. He wore a westerner's business attire: slacks and a sport coat, a white shirt, and a cowboy bolo tie.

At the witness box Frank Craighead raised his right hand as the clerk administered the oath, then took his seat. To his left, Judge Hauk was poring over his résumé, which Stephen Zetterberg had submitted in order to qualify Craighead as an expert witness. *Witness,* however, was an inadequate term for what Craighead was doing there that day. He was as aggrieved as the Walkers were, and he had come seeking justice. Not in the form of money, but rather the redemption of his reputation and that of his twin brother, John, from the injuries done to them by the US government.

The Craighead brothers were at that time America's most famous wildlife scientists. Their outdoor adventures had thrilled people all over the world through magazine articles, newspaper stories, and prime-time television specials. Within the field of wildlife ecology, the Craigheads were predator biologists. This made them unusual. Other than a handful of teaching positions, the job market for wildlife biologists was in fish and game departments and wildlife refuges, whose constituents—hunters and guides—counted on them to produce game species like deer, elk, ducks, and quail, not predators.

The Craigheads were also modern population ecologists. They didn't want to just watch individual animals or try to count them when and where they could find them, like old-time naturalists. They wanted to learn how a whole population of predators functioned in relation to one another, how they regulated their prey, and how, through their prey, they influenced the plants their prey ate and the land over which they ranged. History would ordain that this ambitious vision occurred to the brothers at what would be, at least for them, a fortunate time: the eve of America's involvement in World War II and the Cold War, when a race for technological supremacy and for total knowledge of what a distant enemy was doing gave birth to the tools it would take to realize it.

Frank and John Craighead's work on predation originated in a boyhood fascination with birds of prey. In a joint doctoral dissertation begun in 1941 at the University of Michigan, they set out to study the activities of every meat-eating bird in a thirty-six-square-mile patch of woods and open fields near Ann Arbor. After their research was interrupted by military service during the war, the brothers returned to broaden it to another cohort of raptors in the valley of Jackson Hole, Wyoming. There, and at their Michigan site, the Craigheads aimed to record everything the birds did—where and how they hunted, what they caught, how they bred, where and how they nested, and the number and survival rate of their young.

The lives of some scientists are fairly sedentary, but the Craigheads were possessed of an almost limitless physicality. In the course of their study of hawks, eagles, owls, and falcons in their Jackson Hole site alone, by their own reckoning they walked over 2,000 miles and climbed 1,200 trees to inspect nests. Instead of just typing up their data, they applied it to maps, as a visual representation of the complex web of relations connecting predators to everything around them.

By 1975, the brothers' fame had peaked with their work on Yellowstone grizzlies, and their vision of omniscient 24/7 surveillance of wild animals had been realized in Frank's development of the first large-animal radio tracking collar, an appliance so quickly embraced by wildlife biologists throughout the world that few of us know where it came from. Nevertheless, the résumé Judge Hauk was studying still listed Frank's first love as his chief research interest: "Population Ecology and Raptor Predation."

"ECOLOGY?—OF RAPTOR PREDATION?" the judge asked, looking down through the thick black frames of his reading glasses at Frank's biographical sketch.

By this time ecology was widely known as a political cause, but Judge Hauk wasn't alone in his uncertainty about exactly what it entailed as a profession. In a scant decade and a half ecology had gone from an obscure science to a popular mania. ECOLOGY NOW! demanded bumper stickers, posters, and buttons. Everyone knew we needed it, but its definition as a field of study remained as esoteric to the average person as it was to the federal judge.

"Ecology would mean the biology, the life history, of birds of prey. Their relationship to prey species," volunteered Frank Craighead.

"I always thought *ecology* meant the natural surroundings," the judge mused. "That is, excluding biological specimens, but including plant life. Plant life and dirt and rocks, that sort of thing. Inanimate objects."

Frank was familiar with this misapprehension. Even professional ecologists tended to focus on soil, sunlight, and plants as the regulators of ecosystems. Plants were, after all, the "primary producers" of all the food energy in those systems. Animals came second in the line for scientific attention, and the study of how predators regulated ecosystems from the top down that had fascinated the Craigheads since boyhood had, for a long time, been in last place.

"Ecology is the study of plants and animals, and their interrelationships, and their relationship to the physical and biological environment," Craighead explained to Judge Hauk. "It's really a study of life."

The judge seemed satisfied and declared Craighead an expert witness.

STEPHEN ZETTERBERG ROSE from his seat at the plaintiff's table and carried his notes to the podium, where he began a line of questioning, the intent of which was to elicit from the biologist the details of how a failed experiment in wildlife management had killed Harry Walker.

"How many grizzly bears have you, yourself, actually participated in trapping?" he asked.

"In the course of our work in Yellowstone we trapped and handled over six hundred grizzlies, and I worked with and handled a good percentage of these," replied the biologist.

"How long did you do this?" asked Judge Hauk.

"Intensively over a period from 1959 to 1968," Craighead replied.

"You were not there at the time of this occurrence in question, in June of 1972?" asked the judge, referring to the month of Harry Walker's death.

"No, I wasn't," answered Craighead with all the matter-of-fact detachment he could muster. "Just prior to that, we had been more or less prevented from carrying out our research."

That Frank and his brother—who by 1971 knew more about grizzlies than anyone else in the world—had their access to their research subjects cut off at the height of their fame was the injury Frank hoped to see righted in the Walker trial. But by then there was no righting it. The seeds of both the Craigheads' remarkable success and their fall lay in their own characters. From boyhood they had been filled with a fire of curiosity about wild animals, and they were driven to satisfy it over all resistance. Their success, and eventually this great loss, was founded on a kind of relentless self-assurance that by the time they reached middle age gave no quarter to anyone, and they regarded no authority greater than themselves.

WHEN FRANK AND John Craighead were born in 1916, the official story was that Frank came first, and, as firstborn, was named after his father, Frank Sr., and John arrived a few minutes later. Although twins enter the world at virtually the same time, some research suggests that their personalities and relationships with each other are as affected by birth order as are those of siblings born years apart. The "older" twin is dominant, the "younger" more accommodating, goes this hypothesis.

As boys, the Craigheads were inseparable. Each often knew what the other was thinking, and they finished each other's sentences. Drawn to the same activities, they were intensely competitive, and sometimes argued violently. Their children and graduate students remember how as adults they would storm around their laboratory, yelling at each other. In college they competed as wrestlers; the pictures they both kept in their homes for the rest of their lives showed the two muscular, strikingly handsome boys in singlets, looking at the camera with a measuring gaze that seemed to say: "Bring it on." Throughout their lives they retained a certain grappler's attitude, viewing a challenge of any kind as a call to their masculine courage and iron will to prevail, more than a warning to proceed with caution.

"They were little guys, they didn't like being picked on, and there were two of them," explained John Craighead's youngest son, Johnny.

Their disputes with each other, however, tended to be settled in one direction. John was the unquestioned leader, and he would hold his ground until Frank gave in. Had Frank not done so, he told his children later in life, nothing would have gotten done. Under the birth-order theory, it should have been the other way. The disparity may have an explanation. When the twins were born, the nurses at the hospital put different-colored wristbands on them in order to tell them apart, just as the Craigheads would later do with wild animals. One day, one of the nurses took the bands off the babies while bathing them. She thought she'd put them back on correctly, but admitted she couldn't be certain. In college, the brothers exploited their identical appearance to cover for each other in class when one or the other had to work on a paper. But by virtue of the possible switch when they were babies, not only their professors but their own wives and children never knew for certain which one was which.

Frank and John Craighead grew up in Chevy Chase, Maryland, a suburb of Washington, DC, where their father was an entomologist for the Forest Service. Chevy Chase lies on the northeast side of the Potomac River, where it runs through the coastal Piedmont. In the 1920s the Potomac corridor was a postindustrial wilderness, with the abandoned Chesapeake and Ohio Canal and its towpath, the ruins of nineteenth-century bridge pilings, the great maples and elms, and hawks and eagles nesting in monster sycamores above water-polished sheets of metamorphic bedrock jutting out from the banks. This was where Frank Craighead Sr. took Frank and John to learn about nature.

One day the boys were walking home along the towpath from fishing with their father, when they were startled by something large flying across the trail to alight in the trees.

"What is it, Dad?" they both asked.

"A barred owl," said their father.

Frank had never thought about owls being that big. The owl turned its head, its eyes assessing the twins. Then it took flight again, disappearing into the woods. Frank Sr. looked around for a nest, then pointed it out to the boys, a dark cavity well up the trunk of a big sycamore below the path, toward the river. He suggested they come back another time with a rope to see what was inside. But the twins begged their father to let them try to climb the tree. After considerable struggling, John made it up to where he could peer into the cavity. It took a while for his eyes to adjust, and when they did he could see several fluffy owlets, four or five feet down in the hollow trunk. Frank and John asked their father to let them take one home as a pet. Frank Sr. refused, but the boys pestered him until he relented. The owls were well out of reach, but Frank handed John one of the fishing rods, and after tying a noose on the end of the line and several failed attempts at using it, John was able to catch one of the little birds. The boys wrapped their treasure in a sweater and took it home. It was the first of many birds of prey they would keep as pets.

In their teens the twins made the Potomac River their own—hunting, fishing, canoeing, and watching raptors with their friends. At fourteen they found a 1920 issue of *National Geographic* magazine containing an article titled "Falconry, the Sport of Kings." It became their instructional manual. Equipping themselves with linemen's spurs and ropes, they made daring climbs of trees and cliffs to steal raptor chicks from nests to train. At the same time they became keen on wildlife photography, acquiring bellows cameras and wind-up movie cameras driven by clock springs. Lacking powerful lenses, they built blinds high in trees from which they spied on eagle's nests, and lowered each other down precipices to set cameras in falcon aeries, using long strings to trigger the shutters. A black-and-white movie taken during this period by one of the boys shows the other—it's hard to say which—climbing hand over hand up a rope on a cliff as an angry hawk swoops and dives at him, raking him with its claws. It was not the only time they were attacked by one of the birds. On another occasion Frank, on a cliff, was hit in the head by a peregrine falcon and had to be rescued by his brother and friends.

The twins pursued their incipient study of raptors with the same relentless ingenuity and intensity that would become hallmarks of their later work. The first birds they succeeded in training were Cooper's hawks. They kept the chicks warm under the kitchen stove and as they grew the brothers taught them to fly and return to the glove by proffering bits of liver, meat, and fish. Over the years that followed they tamed and flew prairie falcons, merlins, and kestrels; red-tailed, red-shouldered, broad-winged, and sharp-shinned hawks; and barn, screech, barred, and great horned owls. Their favorites were peregrine falcons, which can plummet toward the earth at two hundred miles per hour in a stoop, hitting their prey with such force as to instantly kill it—as Frank, who luckily was larger and sturdier than a lot of the peregrines' meals, found out.

The boys sold some of their photographs and bought a black, boxy 1928 Chevrolet sedan to take them farther afield. In search of photos and raptor chicks they traveled to Maryland, Pennsylvania, an island off the coast of Virginia, northern Minnesota, and Ontario, Canada. When they graduated from high school in the spring of 1934, they drove west to Wyoming, where they would later settle. In Jackson Hole they sought out one of their heroes, legendary wildlife biologist Olaus Murie, who lived with his wife, Margaret, in a log cabin along the Snake River. Little did the Craighead twins know how the Murie name would figure in their careers four decades later.

In 1935 the two teenagers went to see an editor at the *National Geographic* magazine's offices in Washington, DC, carrying a bundle of photos and a manuscript. They got an assignment, and their feature "Adventures with Birds of Prey" was published in early 1937. Written jointly, it is notable for its unusual authorial voice, in which they share the pronoun *I*. Their editor apparently prevailed on them to clarify who is speaking, so they used "I (Frank)" and "I (John)." The article became the first of nine in which the Craigheads shared their adventures and research with *National Geographic* readers. Their relationship with the National Geographic Society served them well. The society supported their research with grants and publicity for the rest of their careers.

In 1939, the brothers completed undergraduate degrees at Penn State, entered a master's program in ecology and wildlife management at the University of Michigan, published articles in the *Saturday Evening Post* and the *National Geographic*, and saw their first book, on their falconry adventures, released by a major American publisher, Houghton

Mifflin. Meanwhile, their first *National Geographic* article came to the attention of an avid falconer in India. He was a Rajput prince, the son and brother of maharajas, and close to their age. He wrote to the twins through the *National Geographic,* and a correspondence ensued. In the summer of 1939, during a visit to the United States with his brother, the prince flew to Washington, DC, to stay with the Craigheads in the family's modest three-bedroom home. Things were a little awkward in the beginning. John moved out of his room to give the prince his bed and made a pallet on the floor of Frank's room. The prince refused this arrangement, saying that only people of low caste slept on the floor. The Craighead boys reminded him that the outdoor adventures he'd admired in their *National Geographic* article were accomplished by sleeping outdoors on the ground, and he relented. The three of them formed what would become a lifelong friendship, and the prince invited the Craigheads to come to India. *National Geographic* pitched in with another assignment, travel arrangements, cameras, and film.

In September of 1940, the brothers drove across the United States, stopping in Jackson Hole, where they double-dated two young women, both named Margaret, and went to hear a lecture by the now-aged William Henry Jackson. Then they drove to San Francisco and crossed the Pacific on the *President Cleveland,* weathering a typhoon on the way. Arriving in India, they lived for several months in the prince's palatial mansion in Bhavnagar, dressing in tailored white linen and field khakis, flying the prince's birds, running his trained hunting cheetah, and hunting exotic game, while being waited on by servants, drivers, cooks, maids, gun bearers, and the prince's falconers. A photo from the trip that still hung in John Craighead's home in 2015 showed the brothers standing on either side of the prince, all three of them wearing matching silk robes and gilt-threaded turbans.

The Craigheads would have liked to have stayed longer, but World War II closed in and they returned to the United States in February 1941. They brought with them photographs and journals for another *National Geographic* article and footage for a movie they presented at one of Washington, DC's largest venues, Constitution Hall. They were not yet twenty-five.

After India the Craigheads stayed on at the University of Michigan for their doctoral work. When the United States entered the war, they volunteered to teach wilderness survival for the Army, but the Navy picked them up, gave them matching uniforms and officers' commissions, and assigned them to develop training to help carrier pilots and sailors survive if they were shot down or their vessels were sunk in the Pacific. The Craigheads set themselves adrift in life rafts to study how to negotiate surf and coral reefs in order to safely land on an island. They foraged for food, drank from coconuts, and interviewed natives about disease and other perils. From this research, for which they received citations from the Secretary of Defense, they authored a US Navy manual, *How to Survive on Land and Sea,* which remained in print more than seventy years later. By the end of the war, as America sized up its new rival for world domination, the Soviet Union, they were teaching wilderness survival to a more hush-hush clientele, the agents of the OSS, which became the CIA.

John married one of the Jackson Hole Margarets, Margaret Smith, the daughter of a Grand Teton park ranger. John and Margaret and Frank and his new wife, Esther, bought land in Jackson Hole, where the two couples constructed identical log cabins side by side in the sagebrush with a spectacular view of the Teton range. Their wives gave birth to their firstborns on the same day, and each family had two sons and a daughter, all born in pairs in the same years.

The brothers spent the remainder of the 1940s completing their doctoral research and doing what they could to make a living. Their World War II island studies became another feature for *National Geographic,* "We Survive on a Pacific Atoll." By 1950 they were teaching survival for the Strategic Air Command, taking men on training missions that involved running white-water rivers in military life rafts through the wilderness of central Idaho while foraging for wild foods.

With the completion of their doctorates in 1952, John became a professor of wildlife biology at the University of Montana, in a cooperative research program funded by the federal government's Bureau of Sport Fisheries. Meanwhile, Frank's experience seems to have built on their earlier work with the military and spy agencies. A family website states that Frank was doing secret research for the government at this time. If this is true, Frank went to the grave without sharing the details with his family or close colleagues. In 1955 he took a job managing a wildlife refuge for the Bureau of Sport Fisheries and Wildlife near Las Vegas, Nevada. From his residence there he could see mushroom clouds from open-air nuclear explosions on neighboring test ranges, and he became interested in measuring nuclear fallout in desert life. Later, his and John's research at Yellowstone and elsewhere was supported by grants from the Atomic Energy Commission and NASA, and Frank's experiences during the 1950s seem to have fostered in him a lifelong interest in the applications of military surveillance and satellite technology in wildlife research.

There was something distinctly backward-looking about the old North American naturalist tradition, which was aligned with Romanticism in its suspicion toward the modern and the civilized. From Thoreau's retirement to the Concord woods and the popular writing of Grey Owl, an early-twentieth-century English immigrant to Canada who wore his hair in braids and dressed like an Indian, through a whole series of popular naturalists such as Joseph Wood Krutch and Sigurd Olson, the purveyors of wilderness lore traded in nostalgia. But by 1959, when Frank Craighead and his brother came to work at Yellowstone, Frank was anything but an anachronist. True, he and his brother built their log cabins right across the Snake River from the cabins of Olaus Murie and his brother, Adolph, but over time the differences in their philosophies separated the Craigheads and the Muries far more than a couple thousand cubic feet per second of fast-moving water. The Muries were suspicious of technology and of any human intervention in the wilderness. Frank, on the other hand, embraced any new technology he thought would help nature in the nuclear era.

If there was any chance of returning to Eden, thought Frank, it would not be accomplished by pretending the West had not been conquered or that modern tools didn't exist.

6

THE BALANCE OF NATURE

OVER A CENTURY before Charles Darwin's observation that plants and animals were "bound together by a web of complex relations," the eighteenth-century Swedish botanist Carl von Linné wrote a seminal essay on that subject. Linné is better known by his Latin pen name, Linneaus, for codifying what had previously been the hodgepodge classification of plants. Any introductory botany student will know that Linnaeus devised a system under which plants were classified according to the characteristics of their flowers, but fewer undergraduates know about his 1749 treatise on ecology, *Oeconomia Naturae* (The Economy of Nature). The word *economy*—with its implications of thrift and numerical precision, which ecologists hoped would justify their field as a hard science—remained attached to ecology for the next 250 years, through the late-1970s publication of Robert E. Ricklefs's widely used textbook, *The Economy of Nature,* and Donald Worster's history of ecology, *Nature's Economy.*

Linnaeus's essay portrayed an orderly natural world in which each organism had its role in an economy of parts forming a harmonious whole. Later, Darwin supplied a vision of how nature filled each job in the economy with the right someone or something to do it. The perception of harmonious order in nature was, of course, much older than either of them. It went back to the ancients, and the feeling of it still seizes us today when we stand at a lonely beach looking at the perfect tracery of wind and water on sand, or admire how every bit of green light filtering down through the canopy of an old-growth forest is exploited by an intricate tapestry of ferns, moss, mushrooms, and newts on the forest floor. It is an intuition long known as the "balance of nature."

Even before ecology existed as a distinct science, there were dissenters on the issue of balance. Whether you saw order or chaos in nature depended at least in part on what part of it you looked at. Order could be found in ornithology, where centuries of observation by amateur bird-watchers revealed remarkable stability in the number of storks, for example, that returned to nest in European villages. On the other hand, to find chaos you had only to look at the insect world. Entomology was full of reckless excesses, such as biblical plagues of crop-eating grasshoppers. "Where is the balance?" wrote the British scientist Alfred Russel Wallace in 1855. "When the locust devastates vast regions, & causes the death of animals & man what is the meaning of saying the balance is preserved. . . . To human apprehension this is no balance but a struggle in which one often exterminates another."

By the turn of the twentieth century, the idea that ecosystems tend toward homeostatic balance had won out. European and American biologists were refining principles of a new science of community relations between living organisms and their physical environment, and the idea of balance was at its center. According to this model, ecology is not static. Communities of organisms change, but they do so in orderly, predictable steps with equally predictable outcomes. In 1916, the American ecologist Frederic Clements codified this vision of ecology in his *Plant Succession: An Analysis of the Development of Vegetation.* Clements described how bare ground is colonized in a series of predictable stages, culminating in a stable and persistent end state, or "climax." On a beach, just out of the reach of the waves, low-growing, creeping wildflowers establish themselves first. Their roots stabilize windblown sand and their dead leaves decay into humus, forming the first organic soil, which prepares the way for the next predictable assemblage of plants, shrubs, and perhaps, finally, a forest. At each stage the number of organisms, the complexity and stability of the community, and the total mass of biological matter increases. The predestined endpoint, a stable "climax" state, persists indefinitely unless disrupted by a fire, a flood, or human activity, in which case the whole stately process begins again and proceeds toward the same known and predictable end.

For the next half century, the perception that each ecosystem has some preordained ideal state, to which it always seeks to return, informed both ecology and attempts to restore nature in national parks. Like a ship pitched to its side by a wave, then righted by the weight of its ballast and iron keel, an ecosystem would straighten itself out once you removed unwanted human influences.

WITH THE FOUNDING of the Ecological Society of America in December of 1915, ecology announced itself as a recognized field. Among the first things its members did was to express concern about the competing roles of national parks as tourist resorts and nature refuges, which, they said, had gravitated far more to accommodating visitors than preserving a biological legacy. In 1917, the society set up the Committee on the Preservation of Natural Conditions for Ecological Study to lobby for greater care for preservation. The chairman was the great animal ecologist Victor E. Shelford.

Unfortunately, under its first and second directors, the Park Service's emphasis continued to be on courting automotive tourists and their legislative representatives by constructing roads and facilities. By the 1920s, beyond the rangers and superintendents, whose interests were chiefly operational, the Park Service's professional support staff was made up increasingly of landscape architects, whose job was not to study nature but to incorporate facilities into it in an attractive manner. Those with a background in natural sciences found employment as naturalists, whose role it was to explain nature and history to the public—and in the Southwestern pueblos, archaeology—at museums and campfire talks and on guided tours. While some park naturalists did carry on biological research, their conclusions had little reach in decision making within the organization. Instead, many decisions about managing nature were of the seat-of-the-pants kind that one Park Service scientist would later call "cowboy biology."

Animals were looked upon as entertainment for park visitors, and bears were not the only entertainers. In the early days, animals were kept in pens and cages, so visitors could be assured of seeing them. At the Buffalo Ranch in the Lamar Valley, cowhands dressed as Indians simulated buffalo stampedes for visitors' enjoyment. Throughout the national parks, exotic fish were planted to promote fishing as a recreational attraction. At Yellowstone, they included rainbow trout native to the Pacific Slope, as well as eastern brook trout, European brown trout, and lake trout. And, in its first decade and a half after its 1916 charter, the Park Service fell into line with a popular crusade begun by its organizational rival, the Forest Service: killing predators.

THE AGENCY THAT eventually became the Forest Service, first known as the Division of Forestry, came into being in 1881 out of the tireless lobbying of a New York medical doctor, Franklin Hough, who became its first chief. Hough worked as a statistician on the census of 1870, and after crunching the numbers on forest economics, in 1873 he presented a paper to the American Association for the Advancement of Science warning that America's wood supply would be gone in a generation if the government didn't act to secure it for the future. Starting out as a landless advisory agency, the Division of Forestry grew in importance with the creation of a series of "forest reserves" beginning in 1891. In 1905 it was reorganized as the US Forest Service with a domain of sixty million acres of timber reserves, now called national forests.

In the bifurcation of progressive conservation into utility and glory, the Forest Service's role was utility: the husbandry of resources for later use. The Park Service's was glory: the preservation of inspiring landscapes as public resorts. However, in practice, the two agencies did some of the same things. Recreational lodges and resorts were built in national forests, and with the growth in automotive tourism in the 1920s, both agencies were putting in roads, trails, and campgrounds, and seeking congressional patronage to pay for them.

One way the Forest Service found to endear itself to constituents was the extermination of predators. The Forest Service was a major provider of grazing leases under which private cattle and sheep could be driven to summer pastures in the mountains. Not only did the agency wish to assure lessees that their livestock would be safe in national forests, but it also wanted to demonstrate that the forests wouldn't become refuges for carnivores that rampaged on neighboring ranches. At the time, foresters also believed that predators slowed the recovery of populations of herbivores like deer and elk, which had been depleted by overshooting.

In 1905, the year of its reorganization, the Forest Service announced an extermination campaign against wolves, coyotes, mountain lions, and grizzlies. Another federal agency, the Department of the Interior's Bureau of the Biological Survey, had been purposed to study the location and distribution of animals that were considered agricultural pests, and the Survey's biologists took up the job of locating predators for the Forest Service to exterminate.

In 1915, Congress appropriated money for the Biological Survey to conduct its own, independent war on predators, which soon far surpassed the Forest Service's. By the following year the Biological Survey reported it had wiped out 11,000 coyotes in California, Oregon, Nevada, Idaho, and Utah. In 1917 the Survey killed 98 wolves, 1,437 coyotes, and 138 bobcats in Wyoming alone. By 1920, in partnership with the states, the agency extended its deadly franchise to eastern meadowlarks accused of eating oats and corn in South Carolina; robins in the cherry orchards of New York; grebes, loons, terns, gulls, bitterns, and three species of herons feeding at fish hatcheries; and mergansers that were supposedly depleting trout streams in Michigan.

In 1918, the year Park Service rangers took over Yellowstone from the Army, Park Service director Stephen Mather sent a memo ordering Yellowstone's acting superintendent to cooperate with the Biological Survey in persecuting predators. Yellowstone's keepers had previously exterminated wolves, cougars, and coyotes when they had time, but they now became part of a coordinated national machinery that ran down cougars with dogs and caught wolves, coyotes, and wolverines in leg-hold traps and shot them or beat them to death and pulled their pups from their dens and smashed them with cudgels and shovels. In 1917 four wolves were killed in Yellowstone. With the new directive, in 1918, thirty-six wolves and twenty-three mountain lions were dispatched.

In the 1920s, professional ecologists looked upon all the reckless manipulation of the national parks, including persecution of predators, with considerable dismay. The Ecological Society of America passed a 1921 resolution condemning the Park Service's introduction of non-native species such as game fish. A similar resolution was passed by the American Association for the Advancement of Science. In 1925, ecologist Charles C. Adams published a report in the AAAS journal on a survey he'd made of western national parks and national forests. Echoing the reasons for the 1917 creation of the ESA Committee on the Preservation of Natural Conditions for Ecological Study, he wrote that with the alteration of biological systems going on everywhere else, there was a growing consensus among scientists about the value of retaining original conditions in national parks. The Park Service needed to take greater care of this priceless baseline, but instead, the agency seemed hell-bent to pack as many tourists into the parks as possible.

By this time the problem was no longer, as it had been in the nineteenth century, that the science of ecology didn't exist, it was that scientists and up-to-date ecological knowledge were for the most part absent in the workforce of the national parks. The decisions the ESA and the AAAS objected to were being made by rangers who lacked proper training. In his 1925 article, Adams pointed out that while university forestry schools were well equipped to turn out graduates trained for the Forest Service's mission of dispensing a sustainable supply of timber and cattle forage, there was no such educational program to prepare candidates for a career in the preservation of national parks. Adams, who was in New York, may not have known it, but such an incubator for national park science had just come into existence. It was on the far coast, at the University of California at Berkeley.

7

BERKELEY

IT WAS NATURAL for the University of California at Berkeley to become the academic arm of the Park Service in the West. In the 1920s Berkeley was within a day's drive of Yosemite, Sequoia, and General Grant National Parks, across the Central Valley in the Sierra Nevada. Further, the Bay Area favored an appreciation of nature. Berkeleyans could travel by ferry across San Francisco Bay to Marin County, where narrow-gauge logging locomotives towed trains of open-air "gravity cars" up what was billed as "The Crookedest Railroad in the World" to the spectacular 2,372-foot summit of Mount Tamalpais. There the cars were released, engineless, each with a brakeman, to coast silently into the cathedral redwoods of Muir Woods National Monument.

Since the Gold Rush, people had come to California to escape the restraints of convention, and the shoreline of San Francisco Bay continued to foster a socially nonconformist and intellectually adventurous milieu. California remained a treasury of wild nature, yet at the same time it was threatened by wave upon wave of commercial speculation. This instilled in young Berkeley biologists a particularly modern sense of how threatened the natural world was. The Sierra Club and the Save-the-Redwoods league, proto-environmental organizations, both started on the coast of California.

Joseph Grinnell was an early-twentieth-century Berkeley scientist who saw the natural world as ephemeral in that way. A distant cousin of George Grinnell, he was one of those people who understood even before the turn of the twentieth century that he had been born into the middle of a biological catastrophe we now call the sixth great extinction. In 1897, the year he got his bachelor's degree from what is now Cal Tech in Pasadena, he documented the last known tracks of a California grizzly in the San Gabriel Mountains behind Claremont, where Stephen Zetterberg would later grow up and practice law. The bear was extinct in California after 1924.

In 1907, the philanthropist and adventurer Annie M. Alexander endowed UC Berkeley with the funds to create a Museum of Vertebrate Zoology and selected Joseph Grinnell as its director. In this post Grinnell dedicated himself to making a record of what was rapidly being lost around him, in rows of jars and oak drawers full of study skins, skulls, photographs, and notes from field surveys. With their biota under greater protection than in the lands around them, the national parks of the Sierra Nevada were a natural place to inventory native species. Among other projects, for example, between 1911 and 1920 Grinnell and his students made a survey of mammals from the western foothills of the Sierra through Yosemite National Park to the far side of the range at Mono Lake, research that would later prove very useful to the Park Service.

There was also considerable academic cross-pollination between the Museum of Vertebrate Zoology and the forestry school at Berkeley's Hilgard Hall, which by the 1920s also housed a Forest Service research lab known as the Pacific Southwest Experiment Station and a western field office of the National Park Service. As a result, young mammologists and foresters were trained in the national parks and in some cases sought employment in them when they graduated, and the two academic programs and two land management agencies in one place created a think-tank atmosphere that fostered informal exchange of ideas and strategies in offices, hallways, and at social functions.

Berkeley's connection with the Park Service went back to the agency's beginnings. When Congress created the National Park Service to take over parks from the Army in 1916, the Secretary of the Interior appointed a Berkeley graduate and mining millionaire, Stephen Mather, as the service's first director. The following year a young man from Oakland named Ansel Hall graduated from Berkeley with a forestry degree, and in 1920 he went to work as a ranger naturalist in Yosemite, where he founded the park's natural history museum. In 1923, Director Mather promoted Hall to chief naturalist of the Park Service, and instead of bringing him to Washington, allowed him to set up an office at Hilgard Hall.

AMONG THE BERKELEY forestry students who came under the influence of Joseph Grinnell was George Melendez Wright. Orphaned when he was very young, Wright and a family fortune were left in the competent care of an aunt in San Francisco. Like the Craigheads, Wright's first fascination with nature was with birds. In high school he was president of the Audubon Club. By the time he graduated from Berkeley in 1925, he had come to share Grinnell's interest in the conservation of wild animals. He had spent the summer of 1924 traveling by auto to western national parks and wildlife areas, where he'd begun looking critically at how native ecosystems were being managed, and where mistakes were being made. Upon his graduation he became a field assistant to Grinnell and then, following Ansel Hall, got a job as an assistant ranger naturalist in Yosemite.

At Yosemite, Wright witnessed firsthand the "cowboy biology" that was damaging the parks' ecosystems. He watched band-tailed pigeons die after consuming poisoned grain set out by park workers to keep ground squirrels from undermining the footings of a barn. He saw robins and blackbirds perish after coming into contact with oil that park workers had spread on wet areas of Yosemite Valley to control mosquitoes. Lion hunters had been combing the Sierra Nevada, and wolves and grizzlies were already gone. In Yosemite Valley, Wright saw overabundant deer stripping the lower vegetation of trees and shrubs. After his tour of 1924, Wright began thinking about making a survey of wildlife conditions in western national parks similar to Charles Adams's, in the course of which he would categorically describe the types of managerial mistakes he saw in Yosemite.

In 1929, another Berkeley graduate, Horace Albright, a lawyer who served as Stephen Mather's assistant and later as superintendent of Yellowstone, succeeded Mather as director. Wright's aunt had died in 1928, leaving him the family fortune, and he proposed his survey idea to Albright, offering to pay all expenses of the trip and the salaries of two assistants. Albright knew a good deal when he saw one, and he gave his approval.

That November, Wright attended a national conference of Park Service naturalists convened by Ansel Hall at UC Berkeley. In the course of the meeting, he heard a young Yellowstone naturalist, Dorr Yeager, give a talk about whether, and if so how, the Park Service should restore the "balance of nature" in national parks.

"This balance was broken long ago and in most of the parks it was so broken that it can never again be reestablished," Yeager told his fellow ranger naturalists. "It is a very serious question whether or not we should attempt to maintain this balance," he continued. "If it is maintained strictly it will mean that insect epidemics will be allowed to run rampant; and that no steps will be taken in combating forest fires, other than those caused by man." Yeager wasn't sure it was practical to push the parks back to their original states. "There is, I believe, a chance in our National Parks not to preserve the original natural balance, but rather to see that the present state is interfered with as little as possible," he said. "If left alone, I believe that the pendulum of nature will partially return to its original place." Wright already had his doubts about this sort of passivity. By the end of his travels in the next two years, he would come down solidly against it, in favor of purposeful restoration of nature.

In the spring of 1930, Wright left Berkeley with his assistants, Ben Thompson and Joseph Dixon, who had worked on Grinnell's Yosemite survey, in a touring car modified into a sort of flatbed truck. The men sat three-across on a front bench seat, behind which, on the flatbed, was mounted a large chest of scientific instruments, camping equipment, and supplies. They made a big circle through the national parks, which they repeated in 1931. Among the patterns they saw from park to park was a population explosion of herbivores. With predators largely wiped out, Wright's team saw deer and elk, previously depleted from overhunting, now exceeding the land's capacity to support them. The desperate animals were eating anything they could get, chewing the bark off trees, overgrazing meadows and shrubs, and suffering mass die-offs during hard winters.

Another pattern Wright and his colleagues noticed was the habituation of bears to human food, the loss of their fear of human beings, and their increasing aggressiveness. A photograph the team took at Yosemite shows a hole torn through the roof of a 1920s car. Wright suggested that the only solution to problems of this type was secure food storage. "Bears are equipped by nature to tear up obstacles to get their food," he wrote. "They will always get their food wherever it is physically possible for them to reach it. Therefore, if man is going to live in close proximity to bears he must protect his property by devices which bears can not break."

Wright, Dixon, and Thompson also saw widespread problems with park boundaries that had been laid out with no attention to animal migrations. At Mount McKinley, Wright reported, caribou protected inside the park were not protected when they migrated outside of it. The same was true of elk at Yellowstone and Rocky Mountain National Parks. Shoulder to shoulder with domestic sheep in their winter range along lower-elevation park boundaries, wild bighorn sheep were exposed to pinkeye and other exotic diseases. At Yosemite, exotic diseases and overhunting had already resulted in the disappearance of bighorns. Wright thought they should be reintroduced.

In their 1932 report to the Park Service, Wright and his colleagues were clearly in disagreement with Yeager's idea that left alone, nature would sort itself out. Wright wrote that securing the boundaries of national parks against poaching—the struggle of the first twenty years at Yellowstone—was merely a first step and would not be enough to save national parks in the long run. Active intervention based on sound science would be required to restore natural conditions inside them. Wright actually liked that national parks had visitors in them. It would have been simple, he thought, to manage wildlife refuges without visitors. But in a world with a growing human population it was far more interesting to use the parks to study how people and wildlife could live together. "At one bold stroke, man has assumed the whole difficult problem in its most complex form" in national parks, wrote Wright.

Albright very much approved of Wright, and in 1934 he brought him to Washington, DC, to head up a new Wildlife Division—in effect, to be the Park Service's first chief scientist. Shortly thereafter, President Franklin D. Roosevelt appointed Wright head of the National Resources Board. He was thirty years old. In 1936, Wright was traveling with a delegation of Park Service officials to look at possible parklands along the US-Mexican border when he and Roger Toll, then superintendent of Yellowstone, were killed in a head-on collision when a car traveling in the opposite direction suffered a blowout and veered into their lane. Wright's advocacy for scientific study as a basis for sound management of national parks lived on in his successors, among them the wildlife biologist brothers Olaus and Adolph Murie.

The Muries were among the first professional wildlife biologists in America, and they were widely known and respected when they began doing studies for the Park Service. In 1933 Adolph surveyed Isle Royale, in Lake Superior, for designation as a national park. In 1937 he began a study of coyotes at Yellowstone, and from 1939 through the 1960s he studied wolves, Dall sheep, caribou, moose, grizzlies, foxes, and eagles at Mount McKinley National Park in Alaska.* Olaus, with his wife Margaret, did caribou research in Alaska in the mid-1920s. He worked on elk and coyotes for the Bureau of the Biological Survey in Jackson Hole and on grizzlies for the Park Service at Yellowstone. In 1935 he cofounded the Wilderness Society, for which he served as president from 1945 to 1957.

The Muries—with a tiny handful of others in the Wildlife Division,

---

* Mount McKinley National Park is now known as Denali National Park.

including Ben Thompson, Joseph Dixon, Lowell Sumner, and Victor Cahalane—succeeded Wright as the biological conscience of the Park Service. However, something of a philosophical divide had been crossed from Wright to the Muries, without anyone really seeing it at the time. Wright believed that when nature was pushed out of whack by human actions, human intervention was often needed to put it back to rights. But over the years that followed, the Muries came increasingly to advocate leaving nature alone.

"In our national parks we wish to allow nature to take its course as much as possible; we prefer not to carry on any management of the wildlife. We are trying to let alone, to not interfere," wrote Adolph in a 1935 memo to Ben Thompson, who succeeded Wright as chief of the Wildlife Division. And during the pro-conservation administration of Franklin D. Roosevelt the Muries had the ear of Park Service managers. A 1939 policy memorandum by Director Arno Cammerer reiterated Murie's position: "Every species shall be left to carry on its existence unaided . . . unless there is real cause to believe that it will perish if unassisted."

8

CLAYPOOL

CLEARLY, THE ORGANIZED feeding of bears was not in conformance with Park Service director Cammerer's 1939 order that "every species shall be left to carry on its struggle for existence unaided." In 1941, Cammerer's successor, Newton Drury, closed the last bear show in the national parks, at Otter Creek, near Canyon Village, in Yellowstone National Park. But in a lesson the Park Service would have to learn repeatedly, when one source of human food was shut off, bears' search for other sources created dangerous conflicts with people. Interventions such as bear shows and dumps, whether desirable or not, created their own anthropogenic sort of stability. Simply removing them could have consequences as drastic as their original introduction.

In the summer of 1942, after Otter Creek closed, rangers tried to keep a lid on the situation by shooting bears that repeatedly showed up in campgrounds or other developed areas looking for food. By August they'd killed forty-one black bears and twenty-two grizzlies, more than twice as many bears as in the whole of 1941. Nevertheless, on August 23, at 1:45 in the morning, Martha Hansen, a forty-five-year-old supervisory nurse from Idaho, was walking from her Old Faithful rental cabin to a nearby restroom when she was attacked and badly mauled by a grizzly. She was taken to a hospital in Mammoth Hot Springs, where she died of her injuries four days later.

Olaus Murie was dispatched to Yellowstone in 1943 to study the problem. His research focused on what bears ate, and he determined that garbage from dumps and trash cans amounted to no more than about 10 percent of their diet. They could do without it, he said. With the bear shows closed, he urged the park service to cut off other sources of human food—garbage and hand feeding by visitors. Park Service officials had been loath to crack down on hand feeding, because the public loved it so much. But the idea that the Park Service somehow owed visitors the kind of close-up look at animals they could get by feeding them was entirely faulty, argued Olaus Murie. Visitors would appreciate animals more if they had to get out of their cars and hike to see them, he wrote.

Still, the problems of having people sleeping in tents in concentrated numbers along roads were not simple. In addition to Old Faithful, Fishing Bridge campground on the north shore of Yellowstone Lake had been the scene of persistent difficulties with bears, and Murie suggested that the Park Service look at electric fencing. The use of electric fences had already been mentioned by Park Service biologist Victor Cahalane in 1939, and turning Fishing Bridge campground into a fortified compound with a fence and a pit around it was discussed until at least 1945, but was never undertaken.

Grizzly attacks occurred again at Old Faithful, in 1948. On July 13 of that year, William Claypool, a twenty-three-year-old Southern California aircraft plant worker on vacation by automobile with his wife and son, arrived to stay the night at Old Faithful campground. Before setting up their tent, the Claypools asked a ranger whether it was safe to sleep on the ground there. The ranger assured them that it was, and the night passed without incident. However, unbeknownst to the Claypools, during the night of July 12 before their arrival, a bear or bears had rampaged through the same campground, attacking and injuring several people. All available rangers had been called out, a hunt ensued, and a small grizzly was shot. Perhaps to avoid alarming the Claypools, the ranger they talked to had said nothing about it.

The Claypools left Old Faithful to tour other areas of the park, returning around 5:00 p.m. on July 15. During their travels they'd seen bears along the park's roads, and Mrs. Claypool again inquired with a ranger whether it was safe to sleep out. Although every ranger at Old

Faithful must have known about the highly unusual multiple attacks three nights before, Mrs. Claypool was again reassured that there was nothing to worry about. At around 1:00 the next morning, a bear tore through the wall of the Claypools' tent and seized William Claypool by the leg, inflicting serious injuries. Others in the campground were also attacked, although none fatally. Again the rangers responded, and several bears were killed.

Back when Martha Hansen was killed in 1942, the laws of the United States did not permit private citizens to sue the federal government. The only way for victims of such incidents to receive compensation was for Congress to pass a "private relief bill" for each case. As time went on, the legislative calendars of the Senate and House of Representatives became clogged with such bills. In July of 1945, an Army Air Forces bomber got lost in the fog over New York and collided with the Empire State Building, killing several people. The survivors were shocked to learn that there was no way to compel payment from the military. In the aftermath, Congress passed the Federal Tort Claims Act of 1946, allowing for such lawsuits, or tort claims, as they are known in the law. Back home in Southern California, William Claypool sued the National Park Service, and a San Diego federal judge ruled in his favor. The judge found that the Park Service had failed to warn the Claypools of unusually dangerous conditions well-known to rangers at the time and awarded Claypool $5,000 for his injuries.

It was *Claypool v. United States* that Stephen Zetterberg chose as precedent for his Walker lawsuit. Zetterberg's theory of the Walker case was that in an attempt to create natural conditions for wildlife at Yellowstone, the Park Service had finally tried to do between 1968 and 1971 what Olaus Murie had recommended in 1943: close open landfills at which bears were feeding and try to secure trash disposal containers, although the latter effort was uneven. In the process the agency had created an unusual level of danger to citizens from hungry bears entering the campgrounds to look for food, just as had happened in 1941 when Otter Creek closed. And in the Walker case, although the Park Service knew about a troubling increase in bear incidents over a period of four years prior to Harry Walker's arrival, the agency had issued no warning to Harry or other members of the public that something unusual was going on at Yellowstone.

ZETTERBERG CONTINUED HIS direct examination of Frank Craighead, directing Craighead to speak of his sense of impending danger before Harry Walker was attacked.

"During the time that you have just testified that you were working in Yellowstone Park, did you have occasion to sleep out of doors?" he asked.

"I slept outdoors many times in the course of the year," replied the biologist. "Sometimes four and five nights at a time, in snow, following grizzlies to their dens. At other times throughout the summer and the spring, I and my brother, as well as other members of our crew, slept out."

"You have slept out in different parts of the park?" Zetterberg continued.

"Many different areas of the park," responded Craighead.

"Have you slept out in places that are not campgrounds?" asked the attorney.

"Most of our work was off of trails, in backcountry," Craighead answered. "And when I was hiking, yes, I'd sleep right out on the ground in a sleeping bag."

"Have you ever actually slept near where you knew grizzly bears were?" asked Zetterberg.

"I've slept quite close to grizzlies at times," answered the biologist. "On one occasion, within probably three hundred yards of a dozen grizzlies. And my students slept out. My children slept out."

Craighead told the court that he'd had no apprehension about sleeping out until the closure of the Yellowstone dumps. However, by the late sixties, he said, he was warning people to avoid Yellowstone campgrounds such as the one at Canyon, on the rim of the Grand Canyon of the Yellowstone, and the one at Old Faithful.

"I certainly wouldn't have slept in them at that time, myself," he said.

BY THE LATE 1950s, the number of recreationists entering the habitat of the last grizzly bears in the Lower Forty-eight gave good cause to worry, not only about a tiny number of unfortunates like Martha Hansen and William Claypool, but about the future of the bears themselves.

Because the last of them took refuge in the remote mountains of the West, the grizzly is often thought of as a montane species, but the high timber of places like Yellowstone was never their favorite habitat. The grizzly's range once covered much of western North America, from the Arctic well south into Mexico, and they favored the same fertile river valleys and open plains as Euro-American farmers. This overlap, the bear's fearsome size and strength—a locked barn door, a chicken house, or the wall of a cabin were mere inconveniences to them—and their appetite for honeycombs, orchard fruits, and spring lambs put grizzlies and civilization on a collision course. While the black bear thrives shoulder to shoulder with people outside the national parks, grizzlies come into dangerous conflict with human beings in populated, mixed-use lands such as farms or suburbs. Still, they were legendarily hard to kill in the early days. Members of the Lewis and Clark expedition encountered one on a beach along the Missouri River in 1805 and put ten balls from their single-shot muzzleloaders in it before it would die. It was the repeating rifle, Frank and John Craighead always said, that marked the grizzly for extinction in the contiguous forty-eight states, had there not been national parks for it to hide in.

In the parks, however, grizzlies were luckier than wolves and cougars. Early park visitors were interested in bears. Bears can stand up and walk like people. Females sometimes sit upright to nurse their cubs like human mothers. Their curiosity can be comical. One winter day, Frank Craighead watched a young grizzly batting with its paws at the clouds of condensation from its own exhalations. So while wolves and cougars were wiped out, grizzlies were given a reprieve in the national parks. By the 1950s, the last wild grizzly bears south of Canada were split into two populations in the northern Rocky Mountains, with little or no genetic interchange: one in and around Yellowstone National

Park, the other at Glacier National Park, in northern Montana, along the Canadian border. Between the two populations there were estimated to be six hundred to a thousand left, but no one knew for certain. The bears were being joined in their final strongholds by more and more people, and no one was scientifically managing the interactions of the two species. In 1938 Yellowstone had under half a million visitors. After World War II annual visitation exceeded three quarters of a million, and by the late 1950s the park was hosting a million and a half visitors annually, yet it had no biological research program and employed not a single research scientist. After the brief Camelot in the 1930s and early '40s, during which Wright and the Muries had the ear of Park Service management, science in the postwar Park Service entered a Sargasso Sea, where biologists floated, forgotten, without the funds or staff to make an impact on park operations. Olaus Murie left the Park Service for the Wilderness Society, and in the early 1950s, at about the same time John Craighead settled down at the University of Montana in Missoula, Adolph Murie was assigned an office at the headquarters of Grand Teton National Park. With extended forays as a scientist-at-large in the West, Northwest, and Alaska, he remained based there for the rest of his life, a veritable voice in the wilderness. Victor Cahalane, now the Park Service's chief scientist, quit in 1955, in protest over his wholly inadequate budget. Ranger naturalists in individual parks tried to do research, but they were busy answering questions in visitors' centers and presenting campfire talks and nature walks, so they relied on attracting academic scientists to study species that needed attention.

IN 1958, YELLOWSTONE'S chief naturalist, David D. Condon, was concerned about the survival of the grizzly in the Lower Forty-eight. Neither he nor anyone else had any idea how many of them were actually left, and some of the most basic facts about them—the average size of their ranges, their birth and mortality rates, the composition of their population, and the location and constructions of their dens—were largely unknown. Condon was well aware of the kind of incidents that led to a grizzly being shot. He had been chased to his car by one while trying to photograph it. But Condon, an acute observer of wildlife whom Adolph Murie had credited with supporting his work on Yellowstone coyotes in 1937–38, retained a deep appreciation for the species. In June of 1958, he took John Craighead to watch grizzlies feeding at a Yellowstone National Park garbage dump, hoping that he and the University of Montana professor might be able to work out an arrangement for a study.

The Trout Creek dump was twenty-five air miles southeast of park headquarters, near the geographic center of Yellowstone. Originally, Yellowstone had smaller dumps throughout the park, but they had been consolidated into a few larger facilities. In addition, there were some town dumps along the park's boundaries, at Gardiner, Cooke City, and West Yellowstone, Montana. Each spring, when bears emerged from hibernation, hotels and campgrounds opened and trucking and disposal of trash and food waste resumed, large numbers of bears congregated to feed at the Trout Creek dump. That June with Dave Condon, John Craighead saw a concentration of grizzlies that would be hard to beat if you wanted to capture and tag a large percentage of the population to follow the bears' life histories. On a later occasion the Craigheads counted eighty-eight grizzlies at the dump at one time. For a predator biologist this was an unparalleled opportunity: a chance to study America's largest and most charismatic carnivore in America's most famous national park. A memorandum of agreement was drawn up under which, in addition to conducting comprehensive research on grizzlies, the Craigheads would produce recommendations on how the Park Service could better conserve them. The project got under way in June of 1959.

When the study began, John Craighead took a wiry, dark-haired University of Montana undergraduate by the name of Maurice Hornocker to meet his new research subjects. It was late in the day when they arrived in Hayden Valley after the five-hour drive from Missoula. To their left, the Yellowstone River meandered through big meadows, mirroring the sky. Craighead turned right onto an unmarked gravel road leading into the sagebrush-covered hills. A couple of miles up, the road crested a rise and descended toward a dusty pit in the gray-tan soil, within which Hornocker saw heaps of paper, cardboard, wooden fruit and vegetable boxes, cans, bottles, wrappers, kitchen waste from hotels, junk metal, and broken furniture. An unoccupied bulldozer was parked nearby.

The two men got out of the car and waited. Dusk settled over the valley. As the first stars became visible, dark shapes materialized on all sides against the pale sage, headed toward Craighead and Hornocker on intersecting paths, like the spokes of a wheel. Hornocker's heart began beating faster and his breath became shallow. They were grizzlies—big ones. They skirted the two men, seemingly uninterested in them, and descended into the pit. Craighead shone his flashlight on them. The silver guard hairs on their shoulder humps rippled as they pawed through the refuse.

Hornocker would spend a lot of time at the Trout Creek dump over the years that followed. The Craigheads taught him to catch grizzlies in Park Service "culvert traps," used since the 1930s to capture and relocate bears that became nuisances around tourist facilities. The traps consisted of a length of spiral road-drainage conduit large enough to hold a grizzly, mounted on a trailer chassis. The trailer-hitch end was closed by steel plate or mesh, and the other end was fitted with a heavy gate on vertical tracks, like a guillotine, that slammed shut when activated by a baited trigger inside. The Craigheads also trained Hornocker to drug bears by using an archery arrow tipped with a hypodermic dart, and later by using a dart rifle or pistol. Once the grizzlies were immobilized, the Craigheads showed Hornocker how to weigh and measure them and record their sex, markings, physical condition, and other data. The brothers then fixed numbered tags and combinations of bright-colored plastic strips to their ears to individually identify them. Hornocker became the grizzly project's first assistant, and as other students came on board, his specialty became the observation of bear behavior. Grizzlies were most active between sunset and sunrise, so Hornocker often worked at Trout Creek alone, at night. Being out there with a flashlight in one hand, a notepad in the other, and the stench of the garbage, the tinkling of cans and bottles, and the roars and growls of the grizzlies as they defended choice morsels, was an experience the young man never forgot. And the project he was lucky enough to join would turn out to be a study of an entirely new order.

WHEN OLAUS MURIE did his work on grizzlies in 1943, he made observations of individual bears and aggregations of them, and he investigated their diet by watching them feed and examining their scat. However, there were limitations to this kind of work. While a biologist of Murie's powers could recognize individual bears by size, weight, physical appearance, and scars, he could not know a whole population of them, nor could he know more than a fraction of the places they went and what they did when they weren't being watched. Bears travel through thick timber much faster than a horse or human being can, they often move at night, and they can be dangerous to follow.

The Muries believed in studying animals by means of old-time shoe leather. They were critical of scientists who flew around in aircraft and of technology and gadgets. They frowned on the unnecessary handling of park animals. They respected the wildness and autonomy of their subjects. As Olaus put it in his report on the grizzly work, "fauna and flora should be subjected to a minimum of disturbance."

But the Craigheads set out to do something far more ambitious than what Olaus Murie had undertaken. They wanted to make a comprehensive population ecology study of the life histories, demography, behavior, and relationships of as large a sample of the Yellowstone population as they could get, over as many years as possible. They embraced cutting-edge technology, and they had no compunctions about handling animals if it could yield information that might conserve a threatened species. Falconry had made them accustomed to extensive manipulation of wild animals. Once, when they were boys, the tail feathers of their Cooper's hawk were broken in a struggle with a rabbit the bird had caught. The hawk could not fly in this condition, so the twins built it a new tail, splicing owl feathers into the broken feather shafts with glue and tiny splints made of dressmaker's pins. In the late 1940s, they studied the survival of Canada Geese chicks by injecting clutches of eggs of wild birds with vegetable dye, so that the goslings, which are normally yellow, emerged from the shell in garish hues of red, blue, and green. Since the turn of the century, biologists had been putting little numbered metal leg bands on birds, and by the 1920s numbered ear tags on wild mammals, in order to see where they migrated. But the animals had to be recaptured or found dead for the numbers to be read. To get around this problem, the Craigheads installed assemblies of bright-colored plastic loops in sixty-four possible color combinations through a hole they punched, fairly harmlessly, through the cartilage of the bears' ears, so that individual grizzlies could be identified at a distance with binoculars without recapturing them.

IN THEIR FIRST season the Craigheads captured and marked thirty grizzlies. They quickly became very proficient at handling bears, and the rangers came to rely on them to pick up animals captured in campgrounds and move them someplace where they wouldn't be a nuisance. For the Craigheads, each of these requests was an opportunity to add to their research population. However, it soon became clear to them, as it had previously to some rangers, that relocated bears rarely remained where they were released. Totally anesthetized and transported a considerable distance, the bears had an uncanny ability to navigate back to where they'd come from. A two-year-old bear the brothers captured in 1960 was drugged and moved by boat across Yellowstone Lake. In three days it found its way at least thirty-one miles back around the shoreline to where it had been captured. Another bear from the same litter of cubs was moved forty air miles north. It was back the next spring.

The Craigheads used the Trout Creek dump as a release site for bears handed over to them by the rangers. There were a number of things they liked about this arrangement. At their request, the Park Service had installed a locked gate on the road into the dump, and they could handle and release bears there, out of public view and without exposing park visitors to danger. Also, because the dump was near the geographic center of Yellowstone, it was as far as you could get from the park's boundaries on all sides and the dangers grizzlies faced outside them. Further, the brothers thought the food available there might hold bears in the area and keep them from wandering back to the places from which they'd been removed. But even there, bears wouldn't stay put. Under cross-examination by Spivak, Frank Craighead left the witness stand and walked over to an easel on which was mounted a map of Yellowstone, to show what happened.

"One grizzly that was released here," he said, pointing to the dump, "homed all the way back to Gardiner." He pointed to the town on Yellowstone's boundary, thirty air miles north of the dump, but much farther on foot, over rough terrain.

"In other words," said the assistant US attorney, "she was captured at Gardiner and released?"

Craighead confirmed this: "In sixty-two hours she was back in this area where she had been captured, in the town of Gardiner."

9

SMITTY

EACH MORNING BEFORE court, Stephen Zetterberg would get up, dress—he liked fine suits, patterned shirts, and whimsical ties—and eat a breakfast of coffee and poached eggs on toast while reading the *Los Angeles Times.* Then he'd drive to his office a couple of blocks from the Claremont colleges, where the streets are named for famous universities. As a loyal Yale alumnus, Zetterberg's first Claremont office had been on Yale Avenue. By 1969, the firm needed more space, and moved to new, modern quarters on Harvard Avenue, where Zetterberg had the interior painted in rich, contrasting colors and hung with bright expressionist art. From there, he would drive to court. Other attorneys arrived in Chryslers, Lincolns, and Cadillacs, but Zetterberg loved European cars. He had a Volvo, but his favorite was a bright yellow Volkswagen Beetle convertible, in which he traveled the Los Angeles freeways with the top down, whenever possible.

One morning early in the trial, Zetterberg produced from his briefcase a small black-and-white photograph glued to a sheet of paper and with a typewritten caption, and asked that it be entered into the record as exhibit KK-13.

"AK—?" the judge asked.

"KK," Zetterberg corrected him. "Number thirteen, which is a photograph taken by myself of a grizzly bear warning sign at Glacier National Park."

"Any objection?" the judge asked William Spivak.

"I don't see the relevancy," answered Spivak.

"Well," said the judge, "I suppose what its purpose is—is to show that the Interior Department and the Park Service do have certain warnings in Glacier which they do not have in Yellowstone."

What wasn't mentioned on the record, but was known to both the judge and Spivak, was that the existence of the sign in the photograph was the result of another lawsuit Zetterberg had filed on behalf of another bear attack victim in Glacier National Park, that in many ways resembled the Claypool case. The victim, a ten-year-old boy, had been mauled nearly to death on a trail on which another man had been attacked only days before. Yet after the first attack, the Park Service had not closed the trail, nor were any warnings posted at the trailhead.

CONTAINING JUST OVER a million acres of the crest of the northern Rocky Mountains on US side of the Canadian border, Glacier National Park is 248 air miles from Yellowstone's north boundary. To the north it is contiguous with a Canadian national park, Waterton Lakes; to the west and south it adjoins the Flathead and Lewis and Clark national forests; and on the east it abuts the Blackfeet reservation, at the western edge of the Great Plains.

George Grinnell, the crusading magazine editor and founder of the Audubon Society, who advocated for protection of Yellowstone, also worked for the establishment of Glacier National Park. However, as was true of Yellowstone's enabling legislation, the 1910 act creating Glacier was backed by a railroad, the Great Northern, as a resort and passenger destination. In a "see America first" campaign, a railroad subsidiary set up to develop the area, the Glacier Park Hotel Company, compared its snowcapped dolomite mountains, ice fields, and fjord-like lakes to the European Alps. The company constructed a series of large, rustic hotels, including a toll-painted Tyrolean alpine lodge at Many Glacier and a couple of European-style chalets in the backcountry, to which guests were transported on horseback.

As was true at Yellowstone, for decades the waste from these lodgings and the park's campgrounds and picnic areas was disposed of in open-pit landfills. Grizzly and black bears fed at these, at unsecured trash containers, and on food stolen from, or given to them by, park visitors. As was also true at Yellowstone, it took decades and much growth in visitation for anything serious to come of this. In 1939, a hiker was injured by a grizzly; another was bitten in his sleeping bag in 1956. Neither of these was fatal.

In June of 1959, Joseph Williams and Robert Winter, two college students working for the summer at Glacier Park's hotels, went hiking up the Altyn Peak Trail, where Williams was attacked from behind by a small, possibly immature grizzly. He was able to get away and run down the mountain, but the grizzly caught up with him, knocked him down, and began mauling him again. Winter grabbed a rock and began hammering on the grizzly's head. The bear released Williams, who ran away again but tripped and tumbled over a ledge. The grizzly went after him again. Winter tried in vain to distract it, but it paid no attention to him, so he ran down the mountain to summon help. When he returned with two armed rangers, Williams was still alive, with the grizzly on top of him. The rangers shot the bear. Williams survived and sued the Park Service, and the case was settled in his favor.

But something had shifted, some line had been crossed in relations between grizzlies and the growing numbers of hikers, and in the increasing frequency of encounters between the two species lay the premonition of something much worse to come.

IN THE SUMMER of 1960, the supervising ranger naturalist at Saint Mary Lake, on the windswept east side of Glacier National Park, was a Californian by the name of Lloyd Parratt. In the winters he taught college biology in Stephen Zetterberg's hometown of Claremont. Each spring he and his wife, Grace, would cram their outdoor gear, their two teenagers, Mark and Monty, and their ten-year-old boy, Smith, whom everyone called "Smitty," into their station wagon and drive to Montana for the summer.

On July 8, 1960, a hiker was attacked and sustained minor injuries when he encountered a female grizzly with two cubs along the five-mile Roes Creek Trail from Saint Mary Lake to Otokomi Lake. The park was said to have about a thousand miles of footpaths, and at that time the rangers had no procedures for recording grizzly activity and closing trouble spots. Rangers counseled visitors that if approached by a grizzly they should climb a tree. If they couldn't get to one, they should lie facedown and play dead while doing their best to protect their faces, windpipes, and the vital organs of their abdomens.

Ten days after the first attack on the Roes Creek Trail, nothing more had been heard about the sow with cubs. Alan Nelson, a young seasonal ranger who worked under Lloyd Parratt, and a coworker, Ed Mazzer, decided to hike up the trail to fish Otokami Lake. Lloyd's son Smitty was crazy about fishing, and the two rangers invited him to come along. At the lake they encountered two Swedish schoolteachers on a day hike, forty-two-year-old Gote Nyhlen and thirty-eight-year-old Brita Noring. Nelson was pleased to find that Nyhlen shared his interest in wildflowers. In the early afternoon the combined party began making its way back down the trail to Saint Mary Lake with Mazzer leading and Nyhlen and Nelson periodically crouching by the side of the trail to identify flowers.

About a mile down from the lake, spaced out along the footpath, the group crossed an open meadow. On the far side the trail bent back into a forest of subalpine fir. As Mazzer entered the forest he saw a female grizzly approaching in the opposite direction, followed by two cubs. Mazzer turned and ran toward his companions.

"Grizzly bear with cubs! Climb a tree!" He yelled.

Mazzer and the two Swedes scattered uphill, and Nelson and Smitty ran downhill with the bear in pursuit. Smitty could hear the grunts of the grizzly as it closed on him and hit him from behind. Ahead, Nelson reached the base of a fir tree and looked back in time to see the bear throw Smitty around like a rag doll, rip off his scalp, then turn him over and mangle his face. Nelson yelled to distract the animal, and the bear abandoned Smitty and charged toward Nelson. Nelson began clambering up the tree. The lower branches on subalpine firs are small and brittle and Nelson was a big, strong man. The limbs broke, he fell partway back down, and the grizzly seized the back of one of his legs and pulled him out of the tree. When he hit the ground, he had the presence of mind to spread his arms and legs and grab onto the grass. He succeeded in keeping the grizzly from turning him over, but the animal bit a large chunk out of one of his thighs and inflicted terrible injuries to his back and shoulder.

Above the trail, Gote Nyhlen and Brita Noring arrived at the base of a tree. Nyhlen froze. Noring screamed at him to climb. He did. Perhaps distracted by the snapping of limbs at Nyhlen and Noring's tree, the bear ceased its attack on Nelson and ran uphill toward the Swedes. When it got to them Brita Noring was still on the ground. The animal stood up on its hind legs, towering over her. She began to climb, and the animal pulled her down, bit and clawed her, and dragged her, screaming, into the woods. In a tree above, Mazzer found the scene too horrible to watch and turned his eyes away. From the other side of the trail, Nelson could hear Noring's screams. Then, for some reason, the sow left the Swedish woman and began circulating among the other victims, sniffing at them. It sat down and watched them for a while, then got up and ambled away, followed by the cubs.

It was quiet in the meadow after the attack. The murmur of Roes Creek drifted up from below the trail. Smitty was the one most gravely hurt. His scalp was torn off, his face horribly mangled, and one of his eyes dangled on what remained of his cheek. The other eye was obscured in the tatters of his face. One of his arms didn't work. Exploring with his other hand, Smitty discovered a jagged bone sticking out of it. He had multiple broken ribs. There was a hole in his chest that made a bloody gurgling sound when he breathed. Somehow he intuitively performed a maneuver paramedics are trained to employ when confronted with these "sucking chest wounds." Feeling around with his able hand, Smitty located a leaf and covered the hole in his chest, sticking it down with blood. He may have saved his own life. The leaf kept air from filling Smitty's chest cavity outside his lungs, which, if not prevented, causes the lungs to collapse, leading to death by suffocation.

After perhaps twenty minutes, Ed Mazzer and Gote Nyhlen descended from their trees to check on the wounded. The bear had last been seen going down the trail in the same direction they would need to go for help, so they decided to both go, spacing their departures a few minutes apart so that if one of them was attacked, the other might make it through. They departed around 4:15 p.m.

Alan Nelson felt cold and dragged himself out into the sun. Not far away, Smitty drifted in a dreamy state he later described as timeless and beyond fear. At first he was blind, but then a small tunnel of blurred vision opened in the upper part of his functioning eye. He tried to get up and walk but fell down. He found he was now lying in the middle of the trail. He believed he was about to die, so he prayed to God for salvation and just in case, he also said goodbye to his family: "Dear Lord, let me live. Goodbye Mark, Monty, Dad, and Mom." He didn't want to be alone when death came, so he got up again and began limping toward Alan Nelson. Nelson heard a clumping sound and lifted his head to see Smitty staggering toward him. The sight of Smitty's face was too much for him.

"Oh, my God," he said.

Nelson beckoned Smitty to lie down and covered his face with a sweatshirt. Cold and shivering with shock, Smitty snuggled up next to the ranger for comfort. Swarms of mosquitoes settled on them. Nelson dug a shirt out of his pack and spread it over both of them, then arranged the backpack as a pillow under Smitty's head. He told Smitty to lie still in case the bear came back. Then they waited.

Around two and a half hours after the attack, both uninjured men made it down the trail and notified the Park Service. Rangers assembled lights, rifles, wheeled litters, blankets, and medical supplies and started up the trail. It was the day after Lloyd and Grace Parratt's wedding anniversary, and they were eating dinner together when the wife of one of the other rangers came to tell them what had happened. Lloyd and

Smitty's older brother, Mark, went out with the second group of what eventually totaled about forty rescuers.

All three victims were still alive when the rangers reached the scene at 8:30 p.m. Arriving with the second group, Lloyd Parratt searched the scene for Smitty. He found his little boy, swathed in bandages and lying still in a litter surrounded by anxious rangers. As Lloyd approached, one of the men ministering to his son turned, grabbed him by the shoulders, and shoved him away from the litter.

"Lloyd, I tell you, as a father and as a friend, do not look at Smitty. He is not good," said the ranger. But Lloyd pushed by him to comfort his son.

AROUND 9:00 P.M., the rescuers began bumping the three victims down the trail on stretchers equipped with a bicycle wheel in the middle. The jolts made Smitty cry out, weakly. He felt sleepy and asked to be taken home to his own bed. His father stayed by him and held his hand. When they got to the road, the boy was given first aid by a doctor, but his condition was way beyond first aid. The closest hospital was forty-two miles north, across the border, in Cardston, Alberta. Rangers telephoned the staff to be ready to receive three trauma patients. A sergeant in the Royal Canadian Mounted Police was dispatched to pick up two pints of O+ blood from Lethbridge, Alberta. Once the blood was on board he started toward Cardston at speeds of over a hundred miles per hour.

Smitty was loaded into a green Park Service ambulance with a nurse from a park clinic and two visitors who had volunteered to give blood if needed. The other victims were transported in another ambulance. The vehicles sped east along the darkened shore of Saint Mary Lake and turned north through the glacial moraines at the western edge of the Great Plains toward Canada. When they arrived at Cardston Hospital, Smitty was ashen blue and barely breathing. The surgeon who performed the initial survey reported that the boy had no detectable blood pressure or radial pulse. Doctors and nurses warmed him and administered oxygen and plasma before any procedures were undertaken. The Mountie came squealing into the parking lot with the precious blood. Transfusions were started.

Teams of surgeons worked all night on the victims. Smitty had lacerations, puncture wounds to the chest, a collapsed lung, multiple fractured ribs, shattered bones in his face and arm, and neurological damage. Some of the broken bones were exposed through his flesh, where they could easily become infected. He had lost a lot of blood. The eye that dangled down his cheek could not be saved. Doctors put the other back roughly where it belonged, surrounded by damaged tissue. If Smitty lived, said the surgeons, he would likely be blind. Somehow, he made it through the night and the next day. After a month in the hospital in Alberta, he was flown to Children's Hospital Los Angeles, where he began a series of reconstructive surgeries that lasted for years. The Park Service made no attempt to compensate the family for medical expenses, and, at the time, Lloyd and Grace Parratt made no attempt to force them to. Stephen Zetterberg would not hear about the case until four years later.

BY 1960, THE year of Smitty's attack, active claims and lawsuits under the Federal Tort Claims Act for injuries and property damage caused by bears in the national park system reached the million-dollar mark for the first time. The director ordered the writing of the first agency-wide bear management policy. The policy called for installation of bear-proof trash cans and a renewed educational program to discourage visitors from feeding animals. Park regulations allowed the magistrates' courts that heard park cases, such as the court at Mammoth Hot Springs, to impose criminal penalties for feeding bears. But the magistrates generally saw the penalties as excessively harsh, considering that the practice had long been a popular attraction and you still could buy postcards in the stores just outside the parks showing people doing it. So enforcement remained lax, and citations were rarely issued. At Yellowstone, backcountry use by hikers and horseback travelers was relatively light, but Glacier, with its well-advertised resemblance to the European Alps, was known as a hiking park, and more and more people were venturing out on popular trails such as the ones to Otokami Lake, Trout Lake, and Granite Park.

IN THEIR SECOND year of work at Yellowstone, the Craigheads added another thirty-seven marked bears to their study. Every bear got a number and most of them were given names, except for those, such as a young, 175-pound female, number 40, who didn't leave a powerful enough impression of their appearances or personalities the first time they were handled. Those that did, like Pegleg, who had a distinctive, stiff-legged walk; Scarface, Cutlip, and the Rip-Nose Sow, who'd sustained injuries in battles with other bears; Bigfoot; Loverboy; the Fifty-Pound Cub; and the Owl-Faced Sow, whose flat countenance and deep-set eyes reminded the brothers of a short-eared owl—were "gradually taking shape as individuals in our minds," Frank would write later.

As was characteristic of the twins' working lives together, John exercised control over nearly every aspect of the grizzly study. He negotiated the agreement with the Park Service through his Montana Cooperative Wildlife Research Unit at the University of Montana, and he signed all correspondence having to do with it "John Craighead" in the great, slanted loops of his signature, nine to twelve typewriter lines high, over the typewritten title "Leader." However, there was one area of their research that John left entirely up to Frank—perhaps because he never thought it would pan out. As it turned out, Frank Craighead's little project became the study's transcendent innovation.

10

TROUT CREEK

JUST AS ADVANCES in optics allowed seventeenth-century scientists to peer outward through the telescope into the cosmos and inward through the microscope into the cell, so did the development of miniaturized electronics for Cold War missiles and satellites propel wildlife science toward electronic surveillance of the secret lives of animals.

In 1940, a twenty-nine-year-old British ornithologist and committed pacifist, David Lack, volunteered to serve his country after witnessing the horrors of the German night bombing of London. He was assigned to a technical assistance bureau, where he went to work on a problem that had plagued British radar operators. Mysterious ghost objects had been appearing on their screens at night, making it hard to distinguish the next wave of German bombers. Lack suspected that the objects were flocks of birds. British radar experts had considered that possibility but dismissed it, believing that birds didn't fly at night. But Lack's suspicion was confirmed when an observer trained a high-powered telescope on one of the mysterious objects during daylight and saw a flock of seabirds known as gannets. After the war, this discovery paved the way for an experiment in the (intentional) use of radar to observe bird migrations.

In the 1950s, as Cold War defense spending rained down on research and development in electronics, and transistors replaced tubes, a small group of American wildlife biologists grew interested in the application of the tough, miniaturized devices being developed in the aerospace field to the tracking and monitoring of animals. One of them was Dwain

Warner at the University of Minnesota, who complained in 1957 that the Russians had monitored the vital signs of a dog aboard a Sputnik spacecraft orbiting the earth, while American scientists had not done so with animals outside their laboratory windows. Warner and his colleague William Marshall received grants from the National Science Foundation to push the field of wildlife surveillance forward by lacing a plot of university land with buried cables, antennas, sensors, and cameras; fitting wild animals with radio transmitters; and attempting to record everything that went on there. They also developed a tiny tracking radio to put on grouse. In 1959, a meeting was held at the Office of Naval Research in Washington, DC, to discuss the possibilities of electronic wildlife tracking. Carrier pigeons and dogs had been used in military missions, and the Navy would soon deploy trained dolphins for mine clearing and harbor security. Putting transponders on animals was in the interests of both communities.

In 1960, a committee was formed within the Wildlife Society, a professional organization of wildlife biologists, to study wildlife telemetry. Among the founding members was Frank Craighead.

By then, the American military was claiming that its radar could track an object the size of a basketball at a range of eighty-eight miles. Two ornithologists, a young Briton, Ian Nisbet, and a Harvard lecturer, William Drury, got permission to follow up David Lack's World War II work on radar and birds at an American radar installation at Cape Cod. The details of the installation were so secret that the government would not allow Drury and Nisbet to be present and observe the operation for more than a few minutes at a time. So they got permission to set up movie cameras and recorded only what was happening on the radar screens, then reviewed endless hours of film.

In September of 1961, Nisbet and Drury identified the radar signature of flocks of black-bellied and semipalmated plovers migrating over the Massachusetts coast at altitudes of up to twenty thousand feet (pilots of unpressurized aircraft are required to wear oxygen masks above fourteen thousand feet). Radio waves had revealed a phenomenon no one had seen before. That same month, at Yellowstone, John and Frank Craighead fitted a grizzly bear with the first practical large-animal radio collar.

THE CRAIGHEADS HAD spent their first season housed in an old ranger station at Canyon. Then the Yellowstone Park Company, which operated the park's hotels, stores, and restaurants, donated an old employee mess hall nearby for the brothers' use as a field laboratory. No more than a half-hour drive north of the Trout Creek dump, it was a simple, wood-framed building, painted green. They called it the Green Lab. Inside were two large dining tables, a kitchen, storage cabinets, and bunks. On September 21, 1961, John, Frank, and their assistants stayed up most of the night there, putting the finishing touches on a radio collar Frank had been working on, first with a Pennsylvania radio hobbyist and later with engineers from the radio division of the Philco Corporation, a Silicon Valley electronics firm with government contracts in missiles and spy satellites.

In the morning, the Craigheads, Maurice Hornocker, and two other assistants drove to Hayden Valley to check a culvert trap they'd set at the dump. Autumn had arrived in earnest. Snow clouds were gathering, elk were bugling, and the aspen leaves had turned bright yellow. When they got to the dump they found the trap sprung. Inside was the young female grizzly the brothers had previously caught and identified as number 40, but not yet named. She now weighed about three hundred pounds. Hornocker immobilized her by shooting her with a dart loaded with Sucostrin, a paralytic agent. Then the team winched open the door of the trap and John Craighead leaned in to inject the conscious but immobile bear with phenobarbital, a tranquilizer. The team dragged her out of the trap and, working as fast as they could, bolted the padded metal band of the collar around her neck. It had begun to snow, and from their vantage point overlooking Hayden Valley, the men could hear the strange and beautiful calls of sandhill cranes and the howling of coyotes, in the distance through the haze of white flakes.

At first the collar didn't seem to be working. One of the graduate students walked some distance up the valley carrying a radio receiver in a small aluminum case that the Philco engineers had built to Frank's specifications. From there the student reported on his two-way radio that the collar's signal was too weak. The wind picked up, and it began snowing sideways. Snow coated the ground. The men were working bare-handed and their fingers were cold. John and Frank fiddled with the collar and its antenna as the bear began to come out of her stupor. The assistant up the valley radioed to ask how much longer it would take. He was cold. Frank radioed back to ask him if the adjustments had made any difference. They hadn't. There was a period of anxiety. They waited for the grizzly to come to, and when she staggered to her feet and the collar was off the ground, the signal strengthened. The bear walked away, disappearing into the whiteness as the receiver emitted a satisfying *beep, beep, beep.* That sound had not a bit of naturalness to it, Frank Craighead would later remark, but it thrilled him like few sounds ever had.

The Craigheads christened bear number 40 "Marian," after the wife of one of the Philco engineers, and allowed her to disappear into the snowstorm. Then they set out to prove their device by finding her again. Traveling on a heading designed to intersect the bear's course according to the receiver they carried, they climbed a slope, then contoured across it through a grove of pines and into an open meadow. They climbed another hill, and another. The beeping grew louder. Then, looking down from a hilltop, they saw Marian in a swale below them. The device worked.

Over the next decade the Craigheads installed radio collars on twenty-four grizzlies, often multiple times, as batteries went dead or collars required expansion as the animals grew. Each collar had a distinctive signal. Fixed antennas and portable receivers enabled twenty-four-hour tracking of grizzlies' movements, making it possible to arrange a rendezvous between a bear and a researcher who wanted to observe its behavior at any time. "We feel as if we now have a wide-screen view of grizzly behavior, whereas previously we had been peering at it through a keyhole," wrote the brothers.

One of the most immediate revelations of the study concerned the size of the grizzlies' home ranges. Like Drury and Nisbet's radar work, Frank's radio collars revealed something that simply could not have been known before them. Early on, repeated fixes on Pegleg's collar showed his home range at over 160 square miles. Later, work using Frank's invention by the Interagency Grizzly Bear Study Team revealed that the average range of male Yellowstone grizzlies was much bigger: over 1,400 square miles. The radio collar led inexorably to the conclusion that three hundred grizzlies could not be conserved within the confines of Yellowstone National Park. Their survival required a coordinated approach with surrounding federal land management agencies and states.

The Craigheads' work was bearing dividends for grizzlies, but by that September of 1961, when Marian got her collar, there was trouble in the park with elk. That trouble would trigger a major discussion of what was natural and what wasn't, and to what extent unnatural, technological means were justified to achieve natural ends in the national parks. It was a very modern discussion, which ought to have been good for the very modern Craigheads. But change is rarely embraced by everyone involved, without a fight.

PART II

NATURAL REGULATION

11

THE BIG KILL

IN JANUARY OF 1962, a little over two and a half years into the grizzly study, John Craighead summoned his student assistant, Maurice Hornocker, to take a drive from Missoula down to Yellowstone. Craighead wanted to observe something really big going on with elk in the Lamar Valley. When Craighead and Hornocker got to the valley, patches of conifer forest on the surrounding mountains looked almost black in the Arctic glare of the low sun over the Mirror Plateau. It had been particularly cold in Yellowstone that winter, with temperatures dipping below –50 degrees F. The cold stabbed like a knife through the seams of a coat, and at the wrists between the sleeves and gloves. It froze the inside of your nose if you inhaled too deeply. An icy crust had formed on the snow, making it hard for wintering elk to paw through it for food. The Lamar Valley seemed a grander and more terrible place than it had, dressed in the green grass of summer.

Usually, at this time of year, the only tracks in the snow away from the road were the wandering trails of elk, but on their arrival, Craighead and Hornocker found the western end of the valley netted with the bulldozer-like tracks of Park Service Weasels—the boxy closed-cab tractors rangers used to get around when most of Yellowstone's roads were unplowed. The tracks went off in various directions from a cluster of Park Service vehicles and aluminum-sided box trucks and tractor-trailer rigs parked along the road. John Craighead pulled over next to them and he and Hornocker got out.

Around them they saw rangers in green winter uniforms, insulated boots, and trooper hats with fake-fur earflaps, some cradling high-powered rifles. In the distance they heard gunshots. In the backs of the box trucks they saw elk carcasses, field-dressed with their heads and feet cut off, frozen, and stacked like cordwood. They were headed for a meat-packing plant in Livingston, and from there would be given away on the Blackfeet reservation in Montana and to other charities. Craighead and Hornocker were witnessing the largest killing of elk since the 1880s, but this time park officials were the ones doing it. Winged fences had been constructed at the junctions of side creeks and the Lamar and Yellowstone rivers, narrowing into gated corrals. Elk were herded into the enclosures, and then the gates were closed. "It was terrible," said one of the rangers who had participated. "You herded them into a pile, and then you shot into the pile until they were all dead." Nevertheless, John had given the killing the full political support of his research unit at the university. It was a sad thing to witness, he said, but it needed to be done.

OF THE 424 animals and birds placed in the care of the Yellowstone rangers by the act of 1872, few would be as much trouble as bears and elk. But it was elk, in particular, that triggered an argument that lasted for decades over how nature worked, and how much engineering was appropriate to repair it once human beings had damaged it. Behind this was another question: Does nature have a centripetal tendency to go back to what it once was on its own, once it has been disrupted? These questions, triggered by the management of elk, formed the background of the argument about grizzlies that wound up in federal court as *Martin v. United States* in 1975.

Elk don't hibernate as bears do, and for elk, the period of the year when the metabolic demands of staying warm, wading through deep snow, and foraging for food are at their greatest is the time when the plants that make up their diet are dormant or buried under snow. Winter is the grim reaper of grazing and browsing animals in the Rocky Mountains, and the higher up the mountains you go, the more severe the conditions are. So in the autumn, elk move downhill to brushy, south-facing slopes and valley bottoms. In the spring, they "follow the green wave"—as wildlife scientists put it—of new plant growth back up into their summer range. "Their biennial migrations are as regular and systematic as those of migratory birds," observed an Army engineer working at Yellowstone in 1915.

The Continental Divide runs generally north-south in the Rocky Mountains, but in Yellowstone it takes a jog to the west, running sideways across the park. Rain falling south of the Madison Plateau will run toward the Snake River, and north of it, down the Yellowstone. Each autumn, early observers saw scattered bands of elk in the high pastures gather into two principal herds, descending the main drainages on either side of the Continental Divide. The southern elk followed the Snake River to Jackson Hole and beyond, while the northern herd made its way into the Yellowstone Valley—in the early days, more than eighty miles north of the park, toward what is now the town of Big Timber, Montana.

Yellowstone was not created as a wildlife refuge; it was a tourist attraction that began functioning like one. As set forth in the act of 1872, within its boundaries were geothermal curiosities and scenery like the Grand Canyon of the Yellowstone, but not elk winter range in the lower Yellowstone, the Snake, and the Madison river valleys. Yellowstone was a rectangle, and the lives of elk didn't manifest in right angles.

By the late 1890s, with soldiers deployed throughout the park and the Yellowstone Protection Act of 1894 providing federal penalties for poaching, elk were coming back in a big way. "Immense herds can be seen in nearly any direction in winter," reported an acting superintendent in 1898. The following year another superintendent observed that elk "are more numerous than any other animal in the park. . . . They are without a doubt rapidly increasing." As elk increased, persecution of native predators removed a check on the population. From 1904 to 1935, 4,352 coyotes, 121 cougars, and 132 wolves were killed in Yellowstone.

Wolves were finished off by 1926, when the last two were trapped in the Lamar Valley, and in the mid-twentieth century, no one saw a mountain lion in Yellowstone for decades.

With effective park protection and the recovery of the elk population, the differences in the legal status of elk inside and outside the park turned the north boundary into a deadly deterrent to migration. With the coming of elk season, hunters arranged themselves in a skirmish line along the boundary near Gardiner. In October of 1919, early snows drove elk down the Yellowstone River and out of the park, where in the words of one report they "suffered from hunters along the boundary line a percentage loss equal to that of a defeated army." Robert Sterling Yard, director of the National Parks Association and a witness to the carnage, estimated six or seven thousand were killed and another two or three thousand maimed by poor marksmanship. Those that got through were made unwelcome farther down the Yellowstone Valley, where they damaged fences and competed for forage with cattle.

Similar things were happening on the south side of the Divide. In the severe winter of 1909–10, ranchers in Jackson Hole spent freezing nights defending their haystacks as elk starved to death by the thousands around them. By spring there were so many dead elk on the floor of Jackson Hole that a resident claimed he could have walked across the valley on their carcasses without touching the ground. Sportsmen concerned about the elk's future and ranchers tired of seeing their cattle compete for browse prodded the federal government to do something. In 1911, the government started acquiring land north of the town of Jackson for what would become the National Elk Refuge. By 1920, between 850 and 1,900 tons of hay were fed to elk there each winter.

Supplemental feeding also commenced in the north, but since there was no elk refuge outside the park on that side of the divide, feeding took place inside it, in the Lamar Valley and the drainages of the Yellowstone and Gardiner rivers. Coupled with the selective effect of the hail of lead at the boundary, this encouraged elk to remain in the park. There, when they weren't being fed, they gnawed at aspen saplings, stripped the bark off mature trees, and chewed on conifer boughs, willows, currant bushes, and sagebrush. The effects of the restless search for sustenance of some twenty to thirty thousand of them quickly became visible in their winter range, known as the Northern Range. As one park naturalist put it, "All summer [elk] eat the cereal, and all winter they eat the box."

In 1911, the Boone and Crockett Club, a group of enlightened hunters with conservationist leanings, appointed a committee to look into the threadbare appearance of the Northern Range. The committee recommended that some of the elk be removed to spare their habitat. Beginning in 1912, park workers began driving them into corrals and shipping them by rail to restore herds that had been shot out elsewhere in the country. However, not enough were removed to stop the population's growth, and by 1914 the northern herd was estimated at thirty-five thousand, a veritable army of them—or to be more exact, the equivalent of two and a half to three Army divisions—laying siege to plant life on the Northern Range in the winter.

Charles Darwin had referred to the role of a particular species in his "web of complex relations" as its "office," but it was George Wright's mentor at Berkeley, Joseph Grinnell, who coined the term we associate with this idea today: a species' *niche.* No two species could occupy the same niche at the same time. If they did, it often resulted in "competitive exclusion" of one or the other. As elk increased at Yellowstone, whitetail deer, pronghorn, bighorn sheep, and beaver declined. There had once been so many beaver in Yellowstone that in 1863 a traveler griped that he'd had considerable difficulty getting through the lowlands of the Madison River Valley, which had been turned into a swamp by beaver dams. Aspen and willows were beavers' favorite foods. The willows and the beaver had a symbiotic relationship: beaver ate willows and built dams with their wood, and willows grew well in wetlands created by beaver dams. But with tens of thousands of elk chewing up whatever they could to survive, willows grew scarcer, as in turn did beaver. In 1921, a surveyor counted twenty-six active beaver colonies in the Northern Range; two years later he found only nine of them occupied. But the chain of consequence didn't stop there: as beavers disappeared, the landscape changed, too. Over time, as their dams washed out, some brushy wetlands became dry meadows. Birds frequented the willows, narrow-leaf cottonwoods, and aspens in these biotically rich wetlands, but a number of species didn't come to dry meadows.

In what was the first true scientific research in Yellowstone's history, between 1928 and 1932, a Forest Service biologist studied the problems on the Northern Range. He estimated that the land had lost half its productivity since 1914. Reduced plant cover had opened the soil to erosion, and an inch to two inches of topsoil had washed away over more than half the area. Exotic weeds such as cheatgrass, and native rabbitbrush, a dryland species that favors poor ground, were taking over as elk selectively exploited more palatable and nutritious plants.

In just over a half century after trapping and relocation commenced in 1912, Yellowstone National Park gave away thousands of elk to thirty-eight American states and two foreign nations, Canada and Argentina. However, in some of the places where they were released, they created the same problem their removal was intended to solve at Yellowstone, and for the same reasons.

In the Colorado Front Range west of Denver, overhunting eliminated elk by the 1870s and wolves were exterminated by 1900. Beginning in 1913, 350 elk were shipped there from Yellowstone. Two years later Congress protected the chiseled alpine scenery and Christmas-tree conifers of the Front Range by creating Rocky Mountain National Park, but with the neighboring Denver basin already settled in towns and farms, in an oversight repeated throughout the national parks, Rocky Mountain included precious little elk and deer winter range.

As the elk population grew, when the animals came down from the high country in the fall, they became pests on neighboring ranches, gathering like locusts around the hay ricks to consume winter feed intended for cattle and horses. Shot or hazed off the ranches and concentrated in the park, they gnawed at aspens and chewed down willows, sagebrush, and currant bushes, just as they had at Yellowstone. In 1930 and 1931, when George Wright, Ben Thompson, and Joseph Dixon visited Rocky Mountain National Park, they found plant life in the park's winter range in serious decline. As elk consumed the more tender and nutritious grassland plants, porcupine grass and rabbitbrush were taking over. Ben Thompson wrote Director Albright, warning that unless the park acquired more winter range or reduced the number of elk, the lower meadows of Rocky Mountain National Park would be ruined.

AT YELLOWSTONE, EVEN with all of the elk shipped to far-flung places, the condition of the Northern Range didn't get better. In 1934 a drought that decreased plant productivity in most of the United States west of the Mississippi made the imbalance between herbivores' appetites and available food even worse. Following the release of Wright's 1932 report, he and Ben Thompson produced a second one in 1935, in which they included dramatic photos of dying Yellowstone aspens girdled by elk, streamside willows browsed to death, and what their caption called a "sagebrush graveyard" of overgrazed shrubs. It was clear to park officials that live-trapping and relocation had been insufficient to prevent further damage. And so, in 1934, in addition to the continued giveaway program and hunting of animals that left the park, Yellowstone rangers who had previously pursued wolves and cougars turned their guns on overstocked elk. Ten years later, rangers at Rocky Mountain National Park started doing the same thing.

But even this action seemed insufficient to reverse the cascading consequences of unchecked population growth, changes in migratory patterns, and the fact that elk were now feeding with casual nonchalance in thickets they might previously have been loath to enter for fear of wolves and cougars. In 1951, Yellowstone's range management specialist estimated that the Northern Range could support only 5,000 elk. In 1955, the herd was more than twice that size. That winter, 6,535 animals were removed from the northern herd; 1,974 of them were shot on the spot.

The Park Service euphemistically referred to the killings as "direct reductions." Call them what you will, they were highly unpopular among hunters in neighboring states, who thought of Yellowstone as a nursery for game that migrated out of the park in the autumn, into their freezers and taxidermy shops. A pattern was established in which the Park Service would do what it had to, defend its actions in public meetings and the newspapers, wait a year or two for the public to latch on to something else to be angry about, and do it again.

By 1961, the northern herd again numbered over eleven thousand. Yellowstone's superintendent, Lemuel "Lon" Garrison, ordered a much larger kill and began circulating among state and local officials and public forums in Montana to justify what he was about to do. The kill would be carried out in the depths of winter. Butchers were hired to field-dress carcasses, and veterinarians would be on hand to take blood samples and check for communicable diseases. The carcasses would be dragged behind Weasels to central processing areas where they would be hung on racks of pine poles to freeze in the winter air before shipment for processing and health certification and donation to the Blackfeet. In all, in the direct herd reductions witnessed by John Craighead and Maurice Hornocker, the Park Service shot 4,309 elk between November 1961 and February 1962.

PREDATION, WROTE FRANK and John Craighead in the mid-1950s, is a "steady and proportionate pressure that tends to lower the persistent increase in prey species, before more drastic and less steadily functioning forces, such as starvation and disease, run rampant." John Craighead believed that without wolves and cougars, Yellowstone would need human help in order that the elks' starvation not cause damage to the land, and in his role as director of the Cooperative Wildlife Research Unit at the University of Montana, he had given superintendent Garrison his public support in carrying it out. As he saw it, the decision was based on the best available science of the time.

However, much of the public took no stock in the scientific underpinnings of herd reductions. Building on a simmering resentment from 1955–56, the Big Kill of 1961–62 triggered widespread rage, particularly among hunters in neighboring states. In their way of thinking, if elk needed killing, private hunters ought to be the ones doing it. The Park Service argued that the hunting of elk that migrated out of Yellowstone had always been part of the attempted solution, but even when added to the park's trapping and giveaway program, it had never been enough to stop the damage. Hunters responded that for this very reason, national parks needed to be opened to hunting.

Stopping hunting and poaching had been something of a great crusade at Yellowstone and the other early national parks. At Yosemite in 1896, Army administrator Lieutenant Colonel S. B. M. Young, who later served as Yellowstone's superintendent, set up military checkpoints at two of the park's entrances and searched everyone going through them. In the first few months he confiscated two hundred firearms. After all the trouble their predecessors had taken to stop it, the sound of gunshots—unless they were doing the shooting—was unacceptable to rangers.

As the furor grew, Montana congressman Arnold Olsen penned an article in *Sports Afield* magazine titled "Yellowstone's Great Elk Slaughter." The International Association of Game, Fish and Conservation Commissioners, which represented wildlife officials and their hunting constituencies in neighboring states, called for the Park Service to be stripped of the authority to manage wildlife and the responsibility to be transferred to the states. Until this was done, the association promised to block creation of any new national parks. Calls and letters came in to senators and congressmen. Three outfitters with permits to guide elk hunters in the national forest along Yellowstone's eastern boundary filed suit in federal court to stop the killing. School children wrote letters pleading for the animals' lives. Superintendent Garrison said he'd received death threats. Congressional hearings convened in Bozeman, Montana, and a bill was introduced in the 87th Congress to legalize public hunting in the national parks. Facing what was now a full-scale public relations crisis, the director of the Park Service kicked the brouhaha upstairs to John F. Kennedy's Secretary of the Interior, Stewart Udall.

Udall dealt with the crisis in a time-honored way: he called for further study of the issue. He asked Congress to delay action on hunting in parks pending the findings of the two blue-ribbon panels he appointed. One, under the auspices of the National Academy of Sciences, would recommend policies for improving scientific management of national parks. The other would make recommendations specifically on the management of wild animals.

In the spring of 1962, Secretary Udall appointed four prominent wildlife scientists to this second committee. To lead it, he picked Dr. A. Starker Leopold, the oldest son of the late conservationist Aldo Leopold. A highly respected forty-eight-year-old former student of the late Joseph Grinnell, Starker Leopold was a professor of wildlife management at Grinnell's Museum of Vertebrate Zoology at UC Berkeley. No one was better suited to the task, but Udall could not have imagined how much further Leopold would go than he had been asked to go.

12

STARKER

"Your honor, we object to this witness testifying," said Stephen Zetterberg to Judge Hauk as Starker Leopold rose from his seat in the audience to take the stand as a defense witness for the government.

Leopold was a solid man of medium height, half Spanish on his mother's side. At sixty-one his hair was still quite black, parted down the middle of his head in an archaic style, and slicked down to either side with pomade. He was not conventionally handsome—he had an unusually broad forehead, hooded eyes, and a prominent nose set in a long, oval face—but he exuded a magnetic mixture of jovial friendliness and self-assured authority. By 1975 he had for a dozen years been the Park Service's most trusted advisor on how to restore the muddled ecology of national parks.

Zetterberg had objected because Spivak was ten days late in delivering to him a summary of what Leopold could be expected to testify, which the judge required each attorney to provide so that his opponent could prepare for cross-examination.

"Well, let's hear his explanation," said Judge Hauk, turning to Spivak.

Spivak complained that whenever Zetterberg faced testimony that might prejudice his case, he tried to exclude it on technical grounds. Judge Hauk sustained Zetterberg's objection, spent several minutes badgering Spivak for violating his order, and compared calendars with Leopold to find a date to bring the professor back after Spivak turned in his paperwork. None could be found, and Judge Hauk relented.

"Let's let him testify." He shrugged. "I don't care."

As Zetterberg had done for Frank Craighead, Spivak had submitted Starker's résumé to qualify him as an expert witness. Starker Leopold was America's most prominent public-policy expert on wildlife and public lands. He'd served as the chief scientist of the Park Service, on three different advisory committees for the Park Service and the Department of the Interior, and as a member of two presidential commissions. He'd been on the boards of the country's top environmental organizations—the Nature Conservancy, the Sierra Club, the National Wildlife Federation, and the Wilderness Society—and his social connections included corporate CEOs, cabinet secretaries, governors, and legislators. He had consulted on wildlife management for the governments of three foreign nations and written the definitive book on the wildlife of a fourth, and by 1975 his former graduate students occupied many of the top posts in wildlife science and ecology.

After what had been, by 1975, two decades of controversy over how to restore natural conditions to the lives and behavior of Yellowstone animals, Spivak had subpoenaed Leopold to defend the government against Zetterberg's charge that Harry Walker was killed by a Park Service scheme implemented in 1968 under a new superintendent, Jack Anderson, to stop the bears' longstanding use of garbage dumps as a supplemental food source and force them back onto their natural foods.

WILLIAM SPIVAK TOOK the podium. "Doctor, on June twenty-fifth, 1972, do you feel that the risk of harm to park visitors from grizzly bears in the general area of the Walker campsite was increased, or decreased?"

"Clearly, the risk of grizzly bears to people was greatest during the period when the dumps were being closed, when the bears were readjusting their food habits," responded Leopold, in his melodious, breathy, smoker's baritone. His manner was relaxed and congenial.

"Doctor," asked Spivak, "have you observed any antagonism on behalf of the Craigheads toward the Park Service?"

"Objection, Your Honor!" said Zetterberg, "This is not material."

Leopold waited for the judge to rule.

"Yes," responded Hauk. "The relationship of witnesses to either party is permissible. For instance Leopold's relationship with the Park Service, the United States Government, and also the Craigheads, on the question of bias, prejudice, interest, and so forth. Overruled."

Leopold answered the question. "It seems to me there is clearly a disagreement and ill feelings between the Craigheads and Mr. Anderson and other members of the park, stemming back to almost the inception of Superintendent Anderson's arrival at the park. And I can't say what relevance this has to the grizzly bear problem, but there certainly is abundant evidence of strained social relations there, if that's what you mean."

"When did Anderson come to the park?" asked the judge.

"Sixty-seven, I believe," responded Leopold.

Spivak asked him whether the tension between the parties had prejudiced the Craigheads' opinion about closing Yellowstone's dumps abruptly or gradually. Leopold didn't answer the question directly. He acknowledged how much the brothers knew about grizzlies, but opined that at the moment the decision was made, it was anyone's guess what would happen if the dumps were closed quickly or slowly.

The Craigheads had wanted the closure to be accomplished over as much as ten years. To smooth the transition, they recommended that elk carcasses—the park certainly had no shortage of those—could be flown by helicopter to remote bait stations in the backcountry, and these and other areas where grizzlies congregated could be closed to visitors, to keep bears and people apart.

"The contrary view," testified Leopold, "was that if you are going to close the dumps and break these bears from the garbage habit, let's break them and terminate this thing now." He had known things would get dicey. In an article he'd published in *Natural History* in 1970, titled "Weaning Grizzly Bears," he suggested some of Yellowstone's campgrounds might have to be shut during the process, as those at Old Faithful already had been, by then.

"This imposed a greater risk in the first year during the closures," he went on, "but much less risk over the next ten years, because young bears were not learning to eat garbage." He added that by the end of 1971, with the two main dumps closed and the chief research biologist of Yellowstone reporting no human injuries by grizzlies that year, he breathed a sigh of relief.

"I considered the risk was over," Leopold told the court.

At the plaintiff's table, Zetterberg took notes and bided his time. Perhaps he wondered how a scientist of Leopold's caliber could have been mixed up in a mess like this. Anyone who knew the slightest thing about Starker and his remarkable family knew that from earliest childhood, his ecological training had been exceptional, and in his story could be seen practically the entire evolution of ecological thinking in the twentieth century.

STARKER LEOPOLD'S EARLIEST memories were of a time when his father, Aldo Leopold, worked at the district headquarters of the Forest Service in Albuquerque, New Mexico, and his parents were renting a house on South Ninth Street, within walking distance of the Rio Grande. When Starker was four or five, his father fashioned a little seat on the handlebars of his bicycle and began taking him on rides up and down the dusty tracks along the river, with guns or fishing rods slung over his shoulder. It was there, with the wobbly crunch of the bicycle tires on the reddish soil, the warm breezes stirring the cottonwoods, the call of ducks over the water, and his father's voice, close to him, explaining the things they passed, that Starker formed his earliest intuitions about nature.

Starker was the first of five children. Aldo loved all of them, but by virtue of Starker's primacy and his later professional inclinations, he became the closest among his siblings to his father. From those first excursions by bicycle, they hunted and fished together for nearly three decades, sharing their thoughts on the trail, in the duck blind, and around camp at night. They camped in the mesquite, pine, and oak woodlands of New Mexico, Arizona, Texas, and northern Mexico, living on cornbread and stews made of whatever they caught. Later, the two of them went bow hunting in Mexico with archery equipment they fashioned themselves. They lived rough and they lived easy, shivering in the rain dripping off their hats and basking in the firelight as they feasted on venison, game birds, and trout. Later, they canoed the glacial lakes of Minnesota and Wisconsin, where they gorged on northern pike. And everywhere they went Aldo studied nature, and Starker studied Aldo. Indeed, the seeds of Starker's lifetime of achievements were planted in the fertile soil of boyhood, during times spent with his gifted father.

By the early 1920s, two, apparently conflicting, ways of caring for wild places—hands-on scientific manipulation of ecosystems to achieve desired results, and hands-off exclusion of human influence—were expressed in the work of Aldo Leopold. And in Aldo's nimble mind—and Starker's, growing up under his father's tutelage—they were complementary, and not in conflict the way they would be at Yellowstone fifty years later.

In the first instance, in 1924 Aldo Leopold published an article in the *Journal of Forestry* demonstrating how anthropogenic changes to ecosystems, once set in motion, could be self-perpetuating, and would require human intervention if they were to be repaired. Back in 1909, when he graduated from the Yale Forest School and came to work for the Forest Service in Arizona and New Mexico, Aldo had been confronted by unhealthy land. Grazing by too many cattle and sheep had left mountain meadows lacerated by erosional gullies, bleeding soil into rivers now choked with silt and jumping their banks in flood. Local people told him that mesquite, manzanita, and juniper were taking over what had been open grasslands. And during his surveys of the national forests, Aldo also noticed a paucity of wild animals.

Aldo was a scientist with a poet's mind. He tended to describe ecological problems in a large-screen interdisciplinary synthesis of evolution, ecology, and the history of human civilization. He went to work investigating the brush invasion, cutting shrubs and counting their annular rings, inspecting old burn scars on snags and stumps, interviewing old-timers, and reading accounts of what the Southwest had looked like in the journals of early explorers. By 1923, he reconstructed a narrative of human-caused ecological change that is still seen as accurate today, which he published in 1924 in the *Journal of Forestry* as "Grass, Brush, Timber, and Fire in Southern Arizona."

Aldo's research showed that the locals were right. No more than forty years before, a lot of the brush and scrubby woodlands had been perennial grasslands. Before the change, the turf's dense root mats had sheltered the soil from monsoonal rains and made it difficult for trees and shrubs to get established. And the grass communicated frequent fires ignited by lightning and Indians to incipient brush and trees. After fires, grass recovered more quickly than woody plants.

Then, in the 1880s, the Southwest was rapidly settled by Euro-Americans. The Apache, who were known to set a lot of fires, were forcibly ejected. New railroads connected livestock operations to national markets, and a grazing boom ensued. By 1900, some 200,000 cattle and 1.7 million sheep were pastured in the drainage of the upper Rio Grande River. Overgrazing and a reduced fire frequency favored brush over grass. The change was self-reinforcing. The deeper roots of shrubs and trees reduced moisture near the soil surface for grass. With less grass, when fires did start, they weren't carried between patches of brush and trees. It was as if someone had flipped a switch. The Southwest's vegetation converted, and it had no tendency to go back without deliberate human intervention.

Yet in the same year he was indicating that ecosystems damaged by human influence required further human influence to restore them, Aldo founded an institution—the federal wilderness area—that, over time, placed over 100 million acres of public lands under protection from human influence so stringent that today Aldo and Starker would be prohibited from entering them on their bicycle.

Aldo Leopold agreed with the idea championed by the Ecological Society of America that landscapes of various kinds ought to be preserved in natural condition. But in his early writings about wilderness he focused, as had Frederick Olmsted, on wilderness's role as a place for modern people to seek solitude and travel by simple means—foot, horseback, or canoe—away from the roar of civilization. The biggest threat Aldo saw to the remaining wilderness was the growing popularity of automobiles and the rapid construction of roads to serve them. "In 1910," he wrote in 1925, "there were six roadless regions in Arizona and New Mexico, ranging in size from half a million to a million acres. . . . Today roads have eliminated all but one."

Theoretically, preserving land in a form that excluded utilitarian use like timber harvesting would have been the role of national parks, not Leopold's Forest Service. However, since national parks had been set up as public resorts, there was no law or policy prohibiting construction of roads, hotels, and other amenities within them.

The notion of the national forests capturing the ethical high ground by designating areas off-limits to roads and structures came out of a collaboration between Aldo Leopold and a young Forest Service landscape architect in Colorado, Arthur Carhart. In 1919, Carhart was dispatched to survey a new road and cabin lots at Trapper Lake, in the White River National Forest, where he was hectored by two local hunting and fishing guides for destroying the place they loved. Carhart had an epiphany, and he proposed to his boss that the lake be designated a roadless area. His supervisor encountered Aldo Leopold at a meeting in New Mexico and told him about Carhart's idea, and Leopold came to Denver to meet with them. To be effective, Leopold thought, such a place had to be much bigger than Trapper Lake. In 1921 the Forest Service approved Carhart's plan for Trapper Lake, and Leopold submitted a proposal for a three-quarters-of-a-million-acre Gila Wilderness in the Gila National Forest in New Mexico. The Forest Service approved his plan in 1924, and the Gila became the first land in the world officially titled "wilderness." The only way to visit it then, and still today, was on foot or horseback.

WHEN STARKER WAS eleven, his father moved the family to Madison, Wisconsin, where he had accepted a job as director of a Forest Service research station. Aldo expected to be able to carry out ecological research there, but the work turned out to be well described by the name of the place—Forest Products Laboratory—consisting, as Starker remembered it, of chores like figuring out what kind of wood made the best airplane propellers. In 1928 Aldo resigned and went to work as an independent consultant on game management. He performed surveys of huntable wildlife in several states for a trade association of hunting arms and ammunition manufacturers, collaborating with a wildlife biologist from the Florida Panhandle, Herbert Stoddard. The work dried up after the crash of 1929, and the Leopolds lived off their savings while Aldo finished writing the world's first college textbook on wildlife management. The gamble paid off. His *Game Management* was published in 1933, and the University of Wisconsin, where he'd showcased his trade in a series of lectures while he was on his own, hired Aldo Leopold as the world's first "professor of game management." The timing was fortuitous, because the university was about to begin an epoch-making ecological project that would further develop Aldo's thinking—and Starker's, as he followed his father—about the scientific restoration of nature.

University officials had been acquiring land along the shore of Madison's Lake Wingra for an arboretum. Aldo joined the faculty just as the plan came to fruition. He was selected as the arboretum's first director and later served as director of research. University arboretums had traditionally been collections of labeled trees and plants arranged by taxonomy or region. What Aldo had in mind was something entirely different, something that responded to the dystopian horror of an ecological catastrophe unfolding to the south and west of Madison: the dust bowl of the 1930s.

Madison had been the northern edge of the great tallgrass prairie, millions of acres of giant bluestem and other grasses standing eight to ten feet tall, intertwined with violet tufts of blazing star, goldenrod, black-eyed Susan, fringed orchids, and hundreds of other plants. In the late nineteenth and early twentieth centuries, the tallgrass was plowed under and the land was planted in wheat and corn with such breathtaking rapidity that within a single human lifetime the prairie went from stretching as far as the eye could see to becoming one of the rarest ecosystems in the world. Then, in the month after the stock market crash of 1929, came drought, and without the deep roots of the sod to hold it, the soil began blowing away from under the withered crops. In May of 1933, the month before Aldo Leopold was hired to teach at Madison, noon in Dodge City, Kansas, looked more like midnight. At Yellowstone, the drought triggered the first mass shooting of elk.

What could ecologists contribute in the face of an environmental disaster of these proportions? Aldo Leopold and his colleagues planned to learn how to rebuild an ecosystem from scratch, the loss of which had been a key element of the dust bowl.

"OUR IDEA, IN a nutshell, is to reconstruct . . . a sample of original Wisconsin," Aldo Leopold told an audience at a dedication ceremony on a hill overlooking what would be the University of Wisconsin–Madison Arboretum on June 17, 1934. The arboretum would contain ten re-created habitats. Of those, a tallgrass prairie that Leopold penciled on a sketch of the project at a planning meeting in 1933 would become the most famous. In another fortunate historical alignment, the election of Franklin D. Roosevelt and his public works and employment initiatives made labor available in the form of a Civilian Conservation Corps camp established on the grounds. CCC boys, botany professors, and students fanned out into the countryside, collecting surviving constituents of the great tallgrass, and bringing the plants back to Madison. What is today known as the Curtis Prairie became the most well-known and longest-running early example of restoration ecology, the prototype for thousands of acres of public and private prairies that have been restored since.

Among the lessons learned at Madison were two that would be prominent in Starker's later work on the restoration of national parks. In the first case, without meaning to, when Aldo and his fellow professors put together the arboretum, they made its sample ecosystems in the image of the larger world around them—a little Wisconsin without predators. As a professor of game management, Aldo Leopold took a special interest in making sure the arboretum contained not only plants but also animals, and he constructed brush shelters for cover and planted food plants to bring in game birds. However, in Madison, there was no possibility of introducing large predators. The arboretum was soon overrun by a population explosion of rabbits.

The second lesson that carried on from the arboretum through Aldo's later work, and became a major campaign for his son, was the usefulness of fire. The old farmland that became the arboretum was infested with exotic bluegrass, and after frustrating attempts at weeding, in 1938 Ted Sperry, a prairie ecologist at the arboretum, got permission to begin burning the tallgrass, as the prairies after which it had been patterned had burned from lightning and Indian fire. The technique worked, encouraging prairie natives and discouraging the exotics, and by 1943, while Sperry was away at war, a former graduate student of Aldo's was burning the prairie three times a year. It is still burned by their successors today.

In 1935, Aldo extended his restoration experiments to his own land, purchasing a worn-out farm on the banks of the Wisconsin River in the next county north of Madison. The place had been so abused during the depression-era economic undoing of its former inhabitants that it "would grow only cockleburs," Aldo told his wife, Estella, after looking at it with a real estate broker.

In other words, for his purposes, it was perfect.

There Aldo, Estella, Starker, and his four brothers and sisters remodeled a chicken coop into a tiny cabin furnished with bunks, a crude kitchen counter, and a stone fireplace, which they affectionately called "the Shack." On weekends and vacations they all lived together there with books, binoculars, notebooks and pencils, hand tools, and Starker's youngest sister's guitar. Oft-told in the legend of Aldo Leopold—who was prominent in life and posthumously very famous—the Shack was one of those idylls of famous people that behind the appearances was actually idyllic. The Leopolds planted hundreds of pine seedlings, tamarack, mountain ash, wild grapes, and wildflowers. They constructed brush shelters and grapevine tangles for habitat and planted food crops for wild animals. Initially, the results were discouraging. Drought killed 95 percent of the pines and all the mountain ash in 1936. But the Leopolds kept going, and eventually the Shack was surrounded by meadows of native wildflowers and verdant mature forest.

It was in this family culture that Starker came of age. He attended high school in Madison and, following his father's interest in soil conservation, earned an undergraduate degree in soils and agronomy at the University of Wisconsin. In 1936 he began graduate work at the Yale School of Forestry, because his father had gone there. But by this time Starker wanted to be a wildlife biologist like his father, not the forester Aldo had been when he graduated from Yale in 1909. Starker knew of Joseph Grinnell's reputation, and he applied to do graduate work at the Museum of Vertebrate Zoology at UC Berkeley. In 1937 Grinnell took him on as a doctoral candidate.

BY THE LATE 1930s, as Aldo Leopold's restoration experiments at the arboretum and his Sauk County farm unfolded, he could see that if scientists were going to restore damaged temperate-zone North American ecosystems, they would need to see how undamaged ones worked. But where could you see an ecosystem with all its parts and processes still intact? To Aldo's trained eye, it was hard to find a place in the United States that met that standard. He found the answer in the summer of 1936, when he and his brother Carl mounted a bow-hunting expedition to the Sierra Madre, in northern Mexico. In the winter of 1937–38 they went back, taking twenty-four-year-old Starker with them. Starker was just about to begin work on his doctorate, and the trip would prove to be one of the most influential experiences of his life.

Just before the Christmas of 1937 Aldo, Starker, and Carl Leopold met in El Paso, acquired visas from the Mexican consulate, and took a train south into Chihuahua. When they arrived in Casas Grandes, the dirt streets were muddy and it was snowing up higher in the mountains. They spent a freezing night in an unheated hotel, and in the morning their guides came to get them. The guides were members of an expatriate colony of polygamist Mormons who had fled persecution in the United States. They loaded the Leopolds and their gear into a flatbed truck and drove them up a rough, precipitous dirt road to the Mormon colony in the mountains. From there the guides packed them on horseback about twenty-five miles over the top of the Sierra Madre, where they established a camp along the Gavilan River.

The highlands of the Continental Divide, from the Gila Wilderness in New Mexico south into the Mexican Sierra Madre, are volcanic mountains and mesas cut by rocky canyons, forested in oak and pine. Mexican jays chatter in the gray oaks both north and south of the border. To the south they are joined by iridescent-green parrots. However, in many ways the landscapes on either side of the border are similar. It is their human history that is so different. North of the border, the US Army expelled the Apache, but in the Sierra Madre the Indians, and later Pancho Villa's revolutionary gunmen, kept road builders, loggers, and cattlemen at bay.

The differences produced by these conditions were striking to a pair of keen-eyed ecologists. Aldo and Starker noticed that the beautiful, long-needled Apache pines in the Sierra Madre grew farther apart than pines on the American side. The park-like openings between them were carpeted in the luxuriant perennial bunchgrasses that been replaced by brush fields north of the border. The sod acted like a sponge, regulating the release of snowmelt and rain into the Rio Gavilan, a gentle trout stream lined with graceful pale-barked sycamores.

Another major difference from the American side was evident in the area's animal ecology. The United States' war on predators had not reached the Sierra Madre. Starker saw big wolf tracks in the sand along the river and many deer carcasses with the telltale signs of a mountain lion kill: the skull pierced by the lion's teeth, and the carcass covered with duff so the cat could return to feed on it later. Starker recorded in his notes that there were lots of predators, yet there were also lots of deer. After Christmas the weather turned dry, making it hard for the hunters to move without leaves and twigs crackling underfoot. Deer jumped from cover and bounded away in every direction; 253 of them in the course of their stay, Starker noted, or about fifteen per day.

The experience of seeing an ecosystem with all its processes intact would inform Starker's thinking for the rest of his life. He was also astonished to see the Mormon guides leaning out from their saddles to toss lighted matches into the grass as they traveled. It became clear that the admirable condition of those mountains was maintained in part by frequent low-burning fires. "It began to dawn on me that fire was a perfectly normal part of that sort of semi-arid country, and might even be an essential part of it," Starker later recalled. "And Dad, who had been brought up in the Forest Service with the tradition always against fire, he began to wonder too."

ON HIS RETURN from Mexico, Starker went to Berkeley to begin his doctorate. He worked briefly under Joseph Grinnell before the great man died of a heart attack that fall and Starker's supervision was taken over by Grinnell's student Alden Miller. At Berkeley, Starker met an art student from San Diego, Elizabeth Weiskotten, and they married and moved to a tiny cabin in the Ozarks, where Starker did his doctoral research on Merriam's turkeys. World War II came, and Starker considered enlisting in the Navy. He wrote to Alden Miller that he had heard of two brothers, the Craigheads from the University of Michigan, who were teaching wilderness survival for the Navy, and he hoped to enlist and join them. It is hard to imagine how this story might have ended had Starker and the Craigheads shared an important mission on tropical islands in the wartime Pacific. But Starker had crippling allergies, the ragweed was in bloom when he went to Kansas City for his physical, and he failed it. He spent the war years completing his doctorate in Missouri and doing biological research for the Pan-American Union out of Mexico City. Returning to the United States in 1946, he was hired as an assistant professor at the Museum of Vertebrate Zoology.

By 1947, when he took up teaching at Berkeley, Starker joined his father in focusing on a problem that had made Aldo a persistent irritant to state fish and game departments, and would later make Starker Secretary Udall's choice to head his advisory board after the Big Kill: overabundant deer and elk. The most famous case had occurred in the 1920s in Arizona. The Kaibab Plateau, a nearly three-quarters-of-a-million-acre highland on the north side of the Grand Canyon, had been designated as a national game preserve in 1906 during the era of progressive conservation. To restore the deer herd, it was closed to hunting, and predator control agents exterminated 781 mountain lions, 30 wolves, 554 bobcats, and 4,849 coyotes there between the time of the preserve's establishment and 1923. In response to these measures, from about four thousand deer in 1906, the Kaibab deer herd ballooned to an estimated hundred thousand. In 1923 the Forest Service proposed to reopen the preserve to hunting, but the plan was called off because of public opposition. A panel of experts who inspected the area reported that the overgrazing there was the worst any of them had ever seen. Then, in the hard winter of 1924, 60 percent of the Kaibab deer starved to death or, in a weakened condition, died from disease.

Aldo Leopold estimated there had been a hundred such cases with the reduction of predators. The earliest he could find occurred in 1880 in Maine. The Upper Midwest was particularly bad. In Minnesota, pines had stopped reproducing as deer ate the seedlings. In Oregon, 17,000 deer starved to death on the North Fork of the John Day River. Fifteen hundred died of starvation in one winter in the Idaho Primitive Area. Dinosaur National Monument had a chronic problem with overabundant deer. At Yellowstone, rangers shot not only elk but bison and pronghorn, and at Yosemite, Sequoia, and Kings Canyon, they periodically shot mule deer.

The Kaibab became a standard part of Aldo's teaching curriculum and was widely adopted by other instructors in the nascent field of wildlife biology. Not everyone who used the Kaibab story was in favor of letting carnivores live, as Aldo came to be. For most of them, the point of the story was that game managers had to regulate herds with hunting to avoid the destruction of habitat. Guns now did the work predators had done.

However, after his 1938 trip into the Sierra Madre with Starker, Aldo was looking beyond hunters' guns. Back in 1909, when he'd started to work for the Forest Service, he had been as convinced as anyone that predator control was a good idea, and he was personally involved in hunting down wolves. In a famous 1944 essay, "Thinking like a Mountain," Aldo described having shot at a pack of wolves with some other Forest Service men and coming upon a mortally wounded female "in time to see a fierce green fire dying in her eyes." With palpable regret, the essay went on to describe what he witnessed in the years that followed:

> Since then I have lived to see state after state extirpate its wolves. I have watched the face of many a newly wolfless mountain, and seen the south-facing slopes wrinkle with a maze of new deer trails. I have seen every edible bush and seedling browsed . . . to death. I have seen every edible tree defoliated to the height of a saddlehorn. Such a mountainside looks as if someone had given God a new pruning shears and forbidden him all other exercise. In the end the starved bones of the hoped-for deer herd, dead of its own too-much, bleach with the bones of the dead sage, or molder under the high-lined Junipers.*

From 1947, Starker taught the Kaibab too, and he followed his father into the fray. He'd been agitating with the California Division of Fish and Game to do something before California's habitat went the way of

* "High-lined" is a term used to describe trees and shrubs whose lower parts have been uniformly eaten off to an even height that herbivores like deer and elk can reach from the ground. Such vegetation has a peculiar, flat-bottomed appearance that has been used as a measure of overgrazing.

the Kaibab. The division appointed him to supervise a series of deer studies, and under his influence California held its first antlerless deer season on Catalina Island, where mule deer had been introduced for recreational hunting and were destroying vegetation. But the idea of reducing deer herds was as perennially unpopular with hunters as were the elk reductions at Yellowstone.

In the spring of 1948, Aldo and Starker had plans to return to the Sierra Madre. In April, Aldo, Estella, and their youngest daughter, also named Estella, spent a few days continuing the restoration work at the Shack. They picked up their next order of two hundred white pines and two hundred red pines from town for planting. At night Aldo wrote in his journal while his wife knitted and his daughter played her guitar. At about 10:30 in the morning on Wednesday, April 21, they saw a column of smoke rising from a neighbor's land. They drove over, and Aldo reconnoitered the blaze. A trash fire had gotten out of control, and Aldo was worried it would burn into a grove of pines he and his family had worked hard to restore. Aldo went out alone with a backpack firefighting pump to wetline the edge of the burn. Eventually firemen arrived from town to help control the blaze. When Aldo didn't return, friends went to look for him. He was found lying on his back with his hands folded over his chest. The pump was standing on the ground next to him. He had apparently felt some distress, taken the pump off and lain down next to it, and died of a heart attack. The fire burned over him, singeing the notebook he always carried in his breast pocket to record ideas and natural-history observations.

That summer of 1948, after burying his father, Starker returned to the Sierra Madre with a small party of fellow scientists, where he hoped to make an intensive study of how an intact ecosystem worked. He was too late. The narrow, precipitous road to the Mormon village had been widened, graded, and ditched, and logging trucks rumbled down it. Hundreds of cattle were loose in the mountains and the perennial grasslands had been eaten down and infected with European weeds. Without grass to hold them, monsoonal rains swelled the Rio Gavilan, and the flooded river was murky. Predators were being hunted down by cattlemen. Starker returned to Berkeley that autumn with a renewed conviction about the need to save remaining wilderness, about the essential role of predators in healthy ecosystems, and—in the aftermath of his father's death while fighting a brushfire—about allowing wildfire to do its work.

13

PROMETHEUS

STARKER AND ELIZABETH Leopold settled in a little house on Boynton Avenue in North Berkeley. Berkeley was a yeasty place after World War II. Ernest Lawrence, who had worked on the atom bomb, was there. The revered Arts and Crafts architect Bernard Maybeck, designer of San Francisco's Palace of Fine Arts and the university's Faculty Club, was still a legendary presence around Berkeley's school of architecture. The leaders of the Sierra Club's wilderness movement, Richard Leonard and David Brower, were there. Brower lived down the street from the UC philosophy professor Edward Rowell and his wife, Margaret, a renowned cellist and a friend of Mstislav Rostropovich and Pablo Casals. She played duets with Albert Einstein's son, a violinist, who lived next door.

Later, someone made a family tree of the generations of professors and their students at the Museum of Vertebrate Zoology. Starker, having been Joseph Grinnell's student and, after Grinnell's death, a student of his student Alden Miller, was in both the second and third generations. But this was just the most obvious sort of map of an intellectual legacy. Great things are usually produced not by individuals in isolation but by gossamer threads of influence extending from colleague to colleague and friend to friend, like Darwin's complex web of relations in nature. And at Berkeley, Starker came into the presence of scientists who were thinking the way he and his father had been thinking about wildfire.

THERE ARE TWO ways in which most people don't wish to die: by being torn apart by a wild animal and by being roasted in flames. These two abject fears from deep in the ape psyche, became, in the American West, bloated government programs, the two-headed dragon that Starker Leopold fought all of his life.

Five years after the Forest Service began its predator control program, the new agency found an even greater moral purpose in the catastrophic fires in 1910 in Idaho and Montana, which burned 3 million acres and killed eighty-seven people, among them Forest Service rangers who were prominent in fighting them. A political moment was created for the Forest Service to promote itself as the nation's fire department.

As we now know, trying to wipe out fire in the arid West is about as practical as promoting universal chastity before marriage. However futile, the conviction that fire was a universal malefactor in wildlands turned into a moral crusade for which young men, and later young women, were willing to risk death in flames. Early on, this campaign looked to military organizations and methods. In 1919, the Secretary of Agriculture wrote to the Secretary of War, proposing the use of military aircraft to spot fires. Army planes began fire patrols in 1921. In 1940, the idea of paratroopers was adapted to firefighting, and in the postwar years smokejumpers were based from the Rocky Mountains to Alaska to the Southwest, including detachments at the Gila Wilderness and in West Yellowstone, Montana. World War II bombers were also deployed to drop retardant. After the helicopter proved out in Korea, it too was brought to bear on fire. Both smokejumpers and helitack crews were specialized for deployment in roadless wilderness areas to attack fires started by lightning, a force of nature even more powerful than that which threatened premarital chastity.

The Forest Service was by far the biggest operation in fire, and its policies were adopted by the other agencies. After a 50,000-acre blaze at Glacier National Park in 1926 and another at Sequoia National Park in 1927, the Park Service fell into step with the Forest Service under the coordination of the Forest Protection Board, which oversaw fire efforts by federal agencies and the states. The Park Service also hired a chief forester to oversee fire protection in the national parks and installed him with Ansel Hall, the chief naturalist, at the Park Services's western field office in Hilgard Hall at UC Berkeley. In 1933 the Forest Service implemented a "Ten A.M. Policy," decreeing that every fire would be controlled by ten o'clock in the morning on the day after it was reported. This was often impossible, but supplied with enough money to try it, firefighting became a bureaucratic juggernaut, like predator control.

By the time Starker Leopold was teaching at Berkeley, opposition to this machine had coalesced into two tiny groups of insurgent scientists. One gathered at a private research institution near Tallahassee, Florida, around Herbert Stoddard, Starker's father's old friend from the game surveys of 1928–29. The other was at Berkeley.

HERBERT STODDARD WAS a self-trained taxidermist and wildlife biologist who had been working as an exhibit preparer at the Milwaukee Public Museum when the Bureau of the Biological Survey hired him to make a study of southern bobwhite quail on the Georgia-Florida state line. His research was to be paid for via subscription by the wealthy owners of private hunting preserves in the Red Hills, who were worried about the decline in the formerly abundant quail. Many of them were northerners whose families had acquired their estates during the postbellum collapse of southern cotton dynasties in the face of record low cotton prices and the panic of 1893. On the advice of local men hired as gamekeepers, the new owners continued a practice settlers had learned from the Indians: controlled burning to keep the land clear of brush for cattle herding and hunting and to reduce the numbers of ticks, chiggers, and venomous snakes. Burning also seemed to work well to maintain habitat for quail.

In the 1920s an association of private timber owners and the Forest Service waged aggressive antifire public relations efforts in the Southeast. Landowners were discouraged from burning. The woods grew brushy, and the owners of hunting preserves saw a decline in quail. Alarmed, they organized to underwrite a study by the Biological Survey, which dispatched Stoddard from Milwaukee in 1924.

Stoddard's quail study was published in 1931, by which time he had spent some long drives talking with Aldo Leopold about wildlife and fire as they traveled around the Midwest for their game studies. Certainly Stoddard—who, Aldo later said, was the greatest wildlife biologist he ever knew—influenced Aldo's thinking about fire. Stoddard's research showed that not only quail but also the Red Hills' stands of longleaf pine—a species that bore a strong resemblance to the fire-adapted Apache pines that Aldo and Starker were to see in the Sierra Madre—required frequent low-intensity fires to thrive. Stoddard stayed on in the Red Hills, starting a consultancy and bringing in other wildlife experts. In 1958 the group founded an independent fire ecology research center based on one of the old plantations, Tall Timbers.

Privately endowed by the plantation owner's fortune, and free of doctrinal entanglement with government or universities, Tall Timbers Research Station, more than any other American institution, was responsible for incubating a nationwide community of fire scientists and spreading their gospel like sparks across fire lines. In 1962, when Starker's appointment to Udall's board was making him think about how to advise the Park Service, Tall Timbers was hosting the first fire ecology conference in the nation. That zeitgeist and his own strong convictions were certainly part of Starker's liberal interpretation of Udall's assignment, which was simply to tell the Park Service what to do about wild animals. As Herbert Stoddard had demonstrated by 1931, wildlife and fire were connected.

AROUND THE OTHER fire insurgency, in Berkeley, the climate of Northern California was characterized by heavy winter rains that supported lush growths of vegetation, and a six-month dry spell that turned all of that growth into tinder-dry fuel. In spite of antiburning propaganda and pressure from the Forest Service, some Northern California timber owners had continued to burn because burning worked, reducing fuels on the ground that might otherwise allow wildfire to climb into the trees. Some ranchers, too, continued the practice to maintain open grasslands, keep brush under control, and improve the nutrition available to livestock by stimulating new, tender growth on vegetation that had evolved to withstand lightning and Indian-set fires since shortly after the last ice age.

Harold Biswell, a professor of range management in the forestry school, had come to Berkeley to teach the same year Starker did, and they became friends. Biswell had witnessed traditional "woods burning" while doing research on grazing and forestry in the Southeast early in his career. He also knew Herbert Stoddard. Biswell began getting hands-on experience in 1949, burning for California ranchers and timber owners. In 1963, he presented a paper at the second fire ecology conference at Tall Timbers. He became such an evangelist on the subject that people at Berkeley came to call him "Harry the Torch." Starker was tremendously impressed by him. The two men became lunchtime companions at the Faculty Club and coconspirators in reintroducing fire to national parks.

AFTER ALDO LEOPOLD'S death in 1948, Starker's work coalesced around the themes of his father's: fire and predators, protecting remaining wilderness, and manipulating damaged ecosystems back to health. In 1952, the influential conservationist Henry Fairfield Osborne Jr. and his New York Zoological Society sponsored Starker and the British naturalist Frank Fraser Darling in a whirlwind survey of the wildlife and wilderness of the Territory of Alaska. They spent the summer watching caribou, moose, and wolves; camping out; and flying in floatplanes. The experience was similar to that which Starker had had in the Sierra Madre, but this time native people were included in the ecological equation. Attending an Inuit town meeting, Starker was particularly impressed with the intelligence of native Alaskan hunters' observations about wildlife. These were not just recreational shooters; their lives depended on this activity.

During the trip, Starker and Darling landed on a lake to visit the camp of two Park Service planners surveying what later became the Arctic National Wildlife Refuge. On his return to Berkeley, Starker became involved in the effort to secure the area as wilderness. In the fall of 1953, he met with the Park Service planners; Richard Leonard, president of the Sierra Club; and Olaus Murie, now president of the Wilderness Society, to strategize for the political battle to save the area from development. By this time the Muries—Olaus, his wife Margaret, and his brother Adolph—had more or less become the official biologists of the wilderness movement.

In 1955, the American wilderness movement came of age as a force to be reckoned with in American politics when the Sierra Club, the Izaak Walton League, and the Wilderness Society defeated a major dam proposed on the Green and Yampa rivers inside Dinosaur National Monument in Colorado. At the time, wilderness advocates' position could have been summarized as: "Save it, and then leave it alone." But although they worked tirelessly to stop logging, road-building, and dams, they were largely silent about what amounted to a huge habitat modification program in the management of existing wilderness areas: fire suppression. By 1955, Starker Leopold was not silent about it. Not at all.

In a speech that year to the Sierra Club's Fourth Biennial Wilderness Conference, Starker told his audience that wilderness areas ought to be managed "to simulate original conditions as closely as possible." His use of the word *simulate* was a sign of things to come from him. There is a difference between simulating something and making it exactly what it was. A simulation was an approximation, where some prerequisites to a precise re-creation—grizzly bears in Yosemite or wolves at Yellowstone—were not available anymore. But even more important, a simulation had a maker. It was not a thing in itself, but rather the product of an intention in the maker for the benefit of a viewer or user, like a flight simulator was for a pilot.

In his speech, Starker went on to say that wilderness could not be simulated or maintained without the use of fire. American grasslands, chaparral, and forests had been "formed or maintained originally with fire as a component." With fire suppression, they could not be preserved.

"Set aside a sugar pine stand to preserve it in perpetuity," he said, "dominance will shift to incense cedar and white fir." "Set aside a block of prairie," he continued. "A hundred years later it will be a stunted oak-hickory forest. As a matter of policy in preserving natural areas we're going to have to accept responsibility for some experimentation and management."

Leopold spoke again at the next Wilderness Conference, in 1957, and this time he likened the coming appreciation of fire's role in maintaining wild ecosystems to the one conservationists were already developing for the role of predators. Just as coyotes and mountain lions had been labeled evil and destructive but now were appreciated and protected in national parks, so too would be fire. Fire would soon be reintroduced to maintain national parks "in something resembling a virgin state," Starker predicted.

Starker's comments may have been interesting and exciting to some members of the Sierra Club, but they apparently weren't to some Park Service officials who were present. From the back of the room came the angry voice of a former superintendent, Harold Bryant. "I can't believe it . . . a son of Aldo Leopold!" he yelled. An argument ensued between Starker, Bryant, and, jumping in on Bryant's side, the Park Service's chief forester in charge of firefighting. Starker later joked that some of his students had to restrain him from getting physical. But this controversy didn't really worry him. He was an independent voice, a tenured professor friendly with people in the Park Service but in no way beholden to them. It was this intellectual independence that he brought to his outspoken report for the Secretary of the Interior.

In the spring of 1963, Secretary Udall was scheduled to appear with Starker at a convention of wildlife biologists in Detroit, to unveil the findings of Leopold's Advisory Board on Wildlife Management in National Parks. However, when Udall got a look at an advance copy of what Starker had written, he canceled his appearance. Then he watched incredulously as the report garnered raves from environmentalists, scientists, and Park Service management. Bruce Kilgore, a former student of Starker's and now the editor of the Sierra Club's glossy national magazine, the *Sierra Club Bulletin,* published it in its entirety with a glowing review.

"Starker, Goddamn you, you hit a home run, and I didn't know it!" Udall told Starker, as Starker remembered it. Two weeks later the Secretary of the Interior made Starker Leopold's white paper the official policy of the National Park Service.

14

OBSERVABLE ARTIFICIALITY IN ANY FORM

AFTER WILLIAM SPIVAK finished his first round of questioning, Stephen Zetterberg came to the podium to cross-examine Starker Leopold. Zetterberg knew better than to attack Starker's stature as a scientist. A witness's professional accomplishments went to his credibility, but what mattered was what you could get him to say in court about his involvement in the matter at hand.

Starker's weakness on the stand was his strength everywhere else in his professional life. He was an astute politician who thought strategically, a decade out, and he valued the people who would help reach his goals too much to risk his relationships with them over lesser disagreements. But it seemed to Zetterberg that Starker had aligned himself with people whose scientific judgment he didn't entirely trust, and in their effort to curry favor with him, they had lied to him about a fatally dangerous situation in the case of Harry Walker.

"Can a bear get over the habit of going over to the garbage pits in one year?" Zetterberg asked Leopold.

"In one year, if he didn't get anything to eat, he would get pretty skinny if he stood there waiting," Leopold answered. "When he doesn't get rewarded he must have to go do something else for a living, so I suppose he is getting over it—kicking the habit."

"This is your view, anyway, at this time," concluded Zetterberg.

"Yes, sir, indeed," answered the professor.

"Would you disagree with the statement that this problem will not be solved overnight?" asked Zetterberg, and then, without waiting for Leopold to answer, he read aloud from a document on the podium in front of him:

"'Bears conditioned by years of human handouts can hardly be expected to abandon their old habits on command.' Now, would you agree with that statement?" He looked at the professor.

"That's right," answered Leopold, "but within a year—"

Zetterberg cut him off sharply: "You know who wrote that statement? You wrote it!"

"Yes, I know," Leopold admitted. "I do."

IN ADDITION TO its chair, Starker Leopold, Udall's 1962 Advisory Board on Wildlife Management included four other eminent wildlife experts: the chair of the department of conservation at the University of Michigan, both a former director and a former assistant director of the US Bureau of Sport Fisheries and Wildlife, and the executive director of the National Wildlife Federation, who had also been the director of the fish and game departments of both Arizona and Colorado. Yet the board's report was written almost entirely by Starker, and has been referred to ever since as the Leopold Report.

Udall had charged the board with evaluating wildlife management in the national parks, intending to get some direction on managing elk. Starker gave it to him: "Direct removal by killing is the most economical and effective way of regulating ungulates within a park," he wrote. Then, having delivered what was asked for, he jumped the assignment's fence and never looked back. Later, he remarked that he had put a lot of things into the document that he'd been thinking about for a long time. The result was no less than a total philosophy for the ecological management of national parks. It was also a strong call for the manipulation of nature.

Echoing George Wright, Leopold asserted that hands-off protection was inadequate to the task of saving the national parks. He called upon the Park Service to "restore or re-create" natural processes and life communities to bring about conditions as close as possible to those that had been seen by the first Euro-American explorers. The effect would be to display "vignettes of primitive America" for the visiting public.

Echoing his speech at the 1955 Wilderness Conference, Leopold qualified this goal by saying that in many cases the result would be a simulation, not precise restoration. Some elements essential to a faithful reproduction of original American landscapes were gone and would not be coming back. In eastern hardwood forests, monster chestnut trees had been wiped out by the chestnut blight, an exotic fungus introduced from Eurasia. The wingbeats of thousands of passenger pigeons would not be heard again in these forests, nor would southern swamps echo with the hammering of the ivory-billed woodpecker. In the absence of wolves, Yellowstone would always have to deal with too many elk, wrote Starker.

In addition to the missing pieces there were unwelcome additions: Eurasian weeds, English sparrows, Dutch elm disease, and white pine blister rust. Further, national parks were too small to function as complete ecosystems; they could only shelter individual populations of plants and animals isolated from one another by large areas of human-altered landscapes in between.

Given all this, Leopold counseled managers of national parks to achieve a "reasonable illusion" of primitive conditions. To accomplish this would require carrying out scientific manipulation of ecosystems, even a sort of aesthetic stage management. If trees now obscured what early explorers would have seen from a popular viewpoint, they might need to be pruned or cut down. If early explorers had seen pronghorn in Jackson Hole, then pronghorn ought to be there today—whatever it took to maintain them. Native plants could be grown in nurseries and transplanted into the field, where they were a key visual element of the natural scene. Back when thousands of bison had roamed the prairies, their wallows—concavities full of rainwater maintained by the huge animals rolling in them to cool off and clean themselves of insects—also became watering holes for other animals and hosted unique associations of plants and invertebrates. With the great buffalo herds gone it might be necessary to use a bulldozer to simulate these pools. The guide for such construction must be the best science available. Further, the effect should be totally convincing to the public. Starker cautioned park managers that in deliberate restoration of natural scenes, "observable artificiality in any form must be minimized and obscured in every possible way."

This was strong stuff in 1963, but none of it rankled Udall enough for the Secretary to cancel his appearance with Starker at the unveiling of the report in Detroit. What did, Starker later said, was "all that stuff I wrote about fire."

"Of the various methods of manipulating vegetation, the controlled use of fire is the most 'natural' and much the cheapest and easiest to apply," Leopold had written. But setting fires alone would not be adequate. Fuels management, involving crews with chainsaws, would be necessary. Decades of fire suppression had allowed dangerous buildups of dead and downed trees, limbs, and brush. These would have to be cut and removed before experimental burns were started. Even more radical was a passage suggesting that natural conditions would not be accomplished merely with safe little controlled burns. Citing a national park on an island in Lake Superior, Leopold wrote, "On Isle Royale, moose range is created by periodic holocausts. . . ." *Holocausts* was not exactly a politically cautious word to use in discussing the desirability of fires.

Udall had not been completely off base when he guessed that parts of the Leopold Report would be anathema to certain constituencies. Among the report's most shrill critics were leaders of the wilderness movement, who looked upon wilderness as the one place where human beings weren't in charge of everything, and wanted it to stay that way.

"This is the most extreme anti-park policy statement I have yet encountered," wrote Adolph Murie of Leopold's call for scientists to manipulate vegetation in order to simulate views of primeval America from park roads. "Is a scene natural when you chop trees down or plant trees! Is this an honest presentation! Do we want to make Disney Lands out of our roadsides!" In a letter to Park Service biologist Richard Prasil, Murie called Leopold's views "contrary to generally accepted wilderness philosophy."

Howard Zahniser, executive secretary of the Wilderness Society and the principal author of the Wilderness Act, which became law the following year, agreed. Responding to the Leopold Report in the spring 1963 issue of the Wilderness Society's magazine, *Living Wilderness,* he wrote: "In the way in which it deals with national park management in general . . . the board's report poses a serious threat to the wilderness within the national park system and indeed to the wilderness concept itself."

The essential feature of wilderness, Zahniser went on to argue, was the ability of nature to operate free of the sort of human tinkering that was reorganizing everything else in the world. "Such tracts of wilderness land should be managed so as to be left unmanaged," insisted Zahniser, and then summarized his position in an aphorism that would become the slogan of this side of the debate to the present day.

"With regard to areas of wilderness we should be guardians not gardeners."

However, within the Park Service, there was a near-universal embrace of what Leopold had written. Park managers had finally been given a unified field theory of modern ecological park management, well written and easy to understand, with practical examples. Over the years that followed, the Park Service set up a special program at Colorado State University to turn out management trainees who would implement the Leopold Report. When questions about managing nature came up, rangers consulted the Leopold Report like Talmudic scholars. And Starker Leopold became the most trusted voice in the scientific management of natural areas in the United States.

15

RECONSTRUCTION

IN EARLY 1963, in advance of the release of the report of Starker Leopold's Advisory Board on Wildlife Management, Yellowstone made a modest adjustment of the northern elk, trapping and giving away 600 and shooting 406 by the end of that January. Whether by virtue of the relatively small kill, the expectation that the report would soon put an end to the issue, or the distraction of the shadow of nuclear annihilation in the aftermath of the Cuban Missile Crisis in late 1962, there was less trouble about it than before.

John Craighead had been talking with Superintendent Lemuel Garrison about beefing up research on elk to justify further management of the population. Garrison approved John's proposal, and the Craigheads' fieldwork began in 1964. Their study was designed to distinguish the various bands of elk beyond the old northern and southern herds, and to document where each group wintered. Again, the plan required the identification and tracking of individual elk. The Craigheads solved this by capturing the animals and fitting them with brightly colored collars displaying distinctive patterns that could be identified at a distance. Over time, in partnership with the Park Service, they marked 1,448 elk in this manner.

AFTER BEING MAULED in Glacier National Park in 1960, Smitty Parratt underwent more than thirty procedures at Children's Hospital Los Angeles, including bone grafts, cosmetic surgery, implantation of a prosthetic eye, reconstruction of a tear duct, and treatment for a life-threatening bone infection. In 1964, Lloyd Parratt quit the Park Service and took a summer job with the Forest Service in California to be closer to his son. Grace had been teaching private piano lessons out of their home, but these ceased so she could devote herself to Smitty's care. The Parratts remortgaged their home and Smitty's doctors discounted their bills, but by 1964 the family teetered on the brink of bankruptcy. Not knowing what else to do, Lloyd went to see a local attorney someone had recommended: a Claremont native, World War II veteran, and Yale Law graduate by the name of Stephen Zetterberg. Zetterberg saw in Lloyd Parratt a father so loyal to his employer that he had declined to sue until it was too late to do so; and in Lloyd's son, a boy grievously injured by a bear that had attacked someone else on the same trail only days before, yet the trail had not been closed. It was a great case, except there could be no case. The statute of limitations had run out. So Zetterberg set out to get Congress to pass a bill allowing an exception.

In 1948, Zetterberg had run against Richard Nixon for a seat in California's Twelfth Congressional District. His spectacular loss, he would later observe glumly, put Nixon on the road to the White House. But Zetterberg remained well connected, serving on the Democratic Central Committee and advising the governor on health care. So he called in some favors to help the Parratts. California Representative Harry Sheppard introduced a private relief bill to grant an exception to the statute of limitations, which cleared the House in August of 1963. A matching bill was marched through the Senate by the majority leader, Democrat Mike Mansfield, of Montana, where the accident had occurred. The Senate passed the Parratts' bill three days before Christmas, and it went to President Lyndon Johnson, who signed it.

In North Dakota, where he taught school in the wintertime, seasonal ranger Alan Nelson had also struggled under mounting medical bills and time lost from work. The Park Service had turned down his administrative claim for assistance on the grounds that he was off duty at the time he was injured. Suddenly the service reversed itself and placed him on workers' compensation. Nelson, whose wife was best friends with Grace Parratt and kept in close touch with her during the winters, thought this might have something to do with Senator Mansfield. Mansfield was speaking in North Dakota, and Nelson pushed his way through the crowd to express his gratitude. "Don't mention it," Mansfield responded, and went on shaking hands.

So it was that, in 1964, Stephen Zetterberg filed suit against the Park Service at the federal courthouse in Los Angeles on behalf of Smitty Parratt. The case was handed to a young lawyer in the US Attorney's Civil Division on the courthouse's seventeenth floor, whose job it was to defend the government from liability: William B. Spivak Jr.

MEANWHILE, AT SEQUOIA and Kings Canyon National Parks in California, a group of university professors was preparing to generate scientific evidence in support of Starker Leopold's call to allow fire to do its work in forests. For their test case they chose not just any forest but one of the grandest on earth.

Although California's coastal redwoods are taller in sheer mass, giant sequoia trees are the largest single-stem plants in the world. They are also among the oldest, with some specimens living more than three thousand years. Growing more than 30 feet thick at the base and more than 270 feet tall, they exist in only 75 relatively small groves strung out along 260 miles on the west side of California's Sierra Nevada range. At a century in age, "young" sequoias can be perfectly cone-shaped, but in maturity they have an irregular appearance: massive at their buttressed base, with candelabra-like limbs, the lowest of which are often well more than a hundred feet off the ground. Typically, the old ones have flattish or split tops, because they stick up above everything else, exposed to high winds and lightning. They grow widely spaced, allowing beams of sunlight to reach the forest floor. Their wood contains bitter chemicals—tannins—making it so rot-resistant and distasteful to insects that when they fall, sequoia trunks can lie intact on the forest floor for well over a century. Their bark is fibrous, cinnamon-colored, and as much as two feet thick. The bark has sound-absorbent properties, and a reverent hush pervades the ancient groves, broken by the calls of Steller's jays and Douglas squirrels. Standing next to a really big sequoia, people report a distinct sense of an ancient, living presence, something like what one might feel while swimming alongside a blue whale.

Older sequoias display commonsense evidence that they have managed okay, thank you, for millennia, without the assistance of firefighters. Sequoia bark is naturally fire-resistant. In 1959, one of the park naturalists began his evening campfire talks by setting up a blowtorch, with the flame playing on a thick chunk of sequoia bark. When he finished his lecture an hour later, the bark would be blackened but not burned all the way through. Even when fire burns through the bark, it often doesn't kill the tree. Healthy old sequoias display multiple fire scars, some of which let into massive, charcoal-lined cavities where a fire burned deep into the their hearts. Their trunks are so structurally "overbuilt" and their wood so rot- and insect-resistant, that even this doesn't faze them.

If fire was not fatal to sequoia groves, by the early 1960s it was becoming clear that the absence of fire might be. Sequoias had stopped reproducing. Their seeds generally wouldn't germinate in the thick duff on the forest floor, and when they did, they didn't survive. At the same time, fire-intolerant, shade-loving species—white firs and incense cedars—were crowding into what had been open glades between the giants, and as they fell or dropped limbs, the dead, dry material, along with brush, was building up on the forest floor. After eighty years of fire suppression, the big trees stood surrounded by piles of fuel like medieval saints about to be burned at the stake. Harold Biswell, who had started doing research in the groves, found an average of twenty-two tons of dry, highly flammable debris per acre—enough to torch trees that had survived fires for millennia.

In 1964, Biswell got permission to try controlled burning just across Sequoia's western boundary at Whitaker Forest, a 320-acre experiment station owned by Berkeley's forestry school. Whitaker Forest featured a large sequoia grove extending up the humpy ridge of Redwood Mountain into Sequoia and Kings Canyon National Parks. In that same year, on the park side of the boundary on Redwood Mountain, Richard Hartesveldt, Thomas Harvey, and Howard Shellhammer—three academic researchers from San Jose State College on a Park Service research permit like the Craigheads—began a comprehensive study of giant sequoia ecology and the big trees' relations with fire.

On both sides of the boundary, the first burns involved cutting and piling dead material, then burning the piles in the autumn, when moisture conditions made it safe. When the snow melted in 1965, the researchers were overjoyed to find their ash piles from the previous year's burns full of germinating giant sequoia seedlings. Sequoia cones were, in part, serotinous, or fire adapted. Once formed, they might remain closed on the tree for twenty years until the heat from a fire caused them to open and drop their seeds into the ashes after the fire passed. And ashes, Biswell and the San Jose State researchers learned, were a perfect seedbed for sequoias, where germinating seeds had ten times the survival rate they had if they fell on bare soil. The other thing the researchers learned was that giant sequoia seedlings needed sunlight. Once fire suppression allowed firs and cedars to crowd in, a self-perpetuating ecological condition might prevail, such as the one Aldo Leopold had seen in the Southwest, where grassland tipped toward brush by overgrazing would not return to grass. Biswell recommended that invading firs and cedars favored by fire suppression be cut out by forestry workers with chainsaws, after which repeated burning could be employed to restore the supremacy of fire-adapted Sequoias.

As the second year of burning commenced at Redwood Mountain, in the flurry of interest the Leopold Report generated on the impacts fire suppression, at Yellowstone, range management specialist Robert Howe thought there might be an alternative explanation for the decline of aspens on the Northern Range, other than elk: fire suppression and Clementsian succession. Under the theory of succession, fire was looked upon as a force that pushed forests backward in their development to an earlier "subclimax" state. Aspens were considered an early "seral" stage in forest development, which if undisturbed by fire or another disaster, would be succeeded by a climax forest of pine and fir.

The plan was to burn an area where pines and firs were crowding out aspens to see if the aspens would grow back. Ignition was scheduled for the autumn of 1965, when cooler temperatures and the first winter precipitation would prevent a conflagration. In early October, the burn was canceled by a snowstorm. Another attempt was made at the end of the month. In gusty winds and moist conditions, the burn crew sprayed 200 gallons of diesel fuel on the timber and tried to ignite it. They got about ten trees to burn, but further attempts were fruitless. This was the first-ever attempt to intentionally set a forest fire at Yellowstone.

AS SOME OF the first modern experiments with controlled burns in western forests moved forward, in the summer of 1965 Stephen Zetterberg traveled to Glacier National Park to prepare Smitty Parratt's case. Alan Nelson had recovered from his injuries and was assigned to show Zetterberg the scene of the attack. With his notebook and camera, Zetterberg followed Nelson up the Roes Creek Trail. At the trailhead he noticed that the Park Service had installed a sign warning hikers about the presence of grizzly bears and took a picture of it, which, after using it for Smitty's case, he would later submit as an exhibit in the Walker trial. On September 14, 1965, Zetterberg filed a demand letter with William Spivak seeking $329,000 in damages.

Smitty's party could not be faulted for having done much to precipitate the 1960 attack. Incidents involving sows with cubs are some of the least preventable. Smitty was just in the wrong place at the wrong time. However, the rangers knew about, and wrote a report on, the less serious attack on the Roes Creek Trail ten days before. In a letter to Smitty's fellow victim, Brita Noring, after the accident, Glacier superintendent Edward Hummel expressed regrets that no action had been taken earlier: "We now wish we had closed the trail after the first accident." Zetterberg quoted Hummel's admission in his demand letter, citing the principle in *Claypool* of giving the public fair warning about unusually dangerous conditions known to park authorities. In 1966, Spivak settled the case out of court, awarding the Parratt family $100,000 for Smitty's medical bills and occupational rehabilitation.

STEPHEN ZETTERBERG'S NEXT national park case came to him from Yellowstone in the same year he settled Smitty's case. Of the 197 moderate to major injuries in the park that year, sixty-eight involved bears—mostly black bears, and mostly when they were being fed. One case involved a tree. On July 2, 1966, a thirty-nine-year-old man arrived at Lewis Lake campground with his wife and daughter. He set up his tent in one of the campsites and was standing inside it when a 300-year-old lodgepole pine picked that moment to fall on the tent, fatally injuring him. His estate sued the government under the Federal Tort Claims Act. An investigation showed that the tree had a large rot cavity visible from the ground. The court ruled that the rangers ought to have seen it and cut the tree down, and awarded the family over $43,000 in damages.

Elsewhere, seven people were injured when they fell into pools of hot water. Of those, five were at Old Faithful. Of the two who were burned elsewhere, one, a boy named Mark Vaughan, from Los Angeles, became Zetterberg's client. The Vaughan family had stopped their car at a turnout along the shore of Yellowstone Lake and walked past a NO FISHING sign through an area of thermal pools to go fishing. The turnout had previously been posted with danger signs, but the Park Service had removed them in favor of a sign saying the area was closed to fishing. The boy backed into a pool of boiling water, suffered second- and third-degree burns from his chin to his feet, and spent weeks in critical care in Salt Lake City. When he got back to Los Angeles, his father wound up in Stephen Zetterberg's office for the same reason Lloyd Parratt did—to get help with medical bills. Again, Zetterberg faced Assistant US Attorney Spivak, and again Spivak settled before trial, with a $56,000 award for the boy.

In the practice of law, there are arcane specialties. There are expert witnesses who work only on cases involving accidents on commercial fishing boats. There are attorneys who become known for working on the failure of breast implants, construction cranes, or high-voltage electrical panels. When an attorney successfully represents more than one client in a novelty case, word gets around. By 1972, Stephen Zetterberg was known as a good lawyer with a small trade in suing the National Park Service.

THAT AUGUST OF 1966, *National Geographic* ran a sixteen-page article by John and Frank Craighead on their Yellowstone work, lavishly illustrated with color photos. Rugged and sun-bronzed in their flannel shirts and jeans, the brothers looked like everyone's idea of manly westerners, like the cowboys in Marlboro cigarette ads. Since their days as Penn State wrestlers their lives were full of vigorous exercise, and at fifty they had the bodies of men half their age. A Hollywood film crew shadowed them that summer for an hour-long National Geographic television special to be broadcast the next year. Having made their own movies since they were boys, the Craigheads had a highly developed sense of how to work in front of a camera. The crew captured their work with the grizzlies and their wholesome, wilderness way of life, chopping firewood outside their Jackson Hole log cabins, their children rock climbing on Blacktail Butte nearby. The Craigheads taught their sons and daughters to be mountaineers, taught them falconry, fly-fishing, rock climbing, backpacking, skiing, and white-water rafting. They taught their sons to wrestle, and the film crew captured a family ritual in which one of the Craigheads' teenagers flew into the camera's frame, tackling his father from behind without warning and then trying to grapple him to the ground. This and other rollicking scenes were set against an upbeat, Aaron Copland–esque symphonic score, like the soundtrack to a western movie. The Craigheads were rock stars. Going into 1967 with that level of public admiration, they may have felt politically bulletproof.

IN THE SPRING of 1967, the Vietnam War came home to Anniston, Alabama, and Oxford, Anniston's neighbor to the south, where Harry Walker went to high school. In 1966, 382,000 young American men were conscripted and, by early 1967, 400,000 were on the ground in Southeast Asia. A hundred thousand more joined them that year, and some of those from Anniston and Oxford didn't make it back. John Wayne Hudgens was shot that spring while serving with the 101st Airborne Division. Donald Barnett, a classmate of Harry's sister Carolyn, burned to death in an attack on the American base at Bien Hoa. William Sapp died in a helicopter crash. Willard Young died from multiple shrapnel wounds. So did Anniston's Jimmy Alexander, in Quang Ngai Province, at the age of twenty. Howard Ray Thomas, also twenty, was killed in action in Long Khanh Province.

Among Wallace Walker's fellow dairymen was a farmer with a ramrod-straight military bearing who'd been an Army officer during the Korean War. Like many small farmers, he supplemented his income with a second and then a third job. In the 1960s he was the state health department's milk inspector for the Choccolocco Valley and, at the same time, commander of the local contingent of the National Guard. In the course of his visits to the dairies, he recruited fellow farmers' sons he knew to be of good character. Guardsmen were exempt from conscription into the regular armed services, and in those days the National Guard was not used overseas. So the effect was to keep boys home to work on their parents' farms.

The milk inspector had a high opinion of Wallace and his family. In 1967, when Harry was without a draft deferment and the military was sucking up poor boys all over Alabama, the inspector suggested that Harry join the guard. Harry signed up, and after basic training he received a modest supplementary paycheck in exchange for a weekend assembly each month and two weeks at a training camp in Mississippi each summer. A skilled hunter, Harry distinguished himself as a marksman and earned a rifleman pin. A family photo from this time shows him standing next to his car outside the Walkers' farmhouse in his dress uniform. He is clean-shaven, his hair is neatly clipped, and he looks very proud.

BY THE FALL of 1966 there were an estimated 7,703 elk on the Northern Range. Yellowstone superintendent John McLaughlin, who had succeeded Lemuel Garrison, made plans to shoot some more. In March of 1967, the rangers were back out in the Lamar Valley and at Slough Creek in their green winter uniforms, calf-high insulated boots, and fur-lined trooper hats, driving around with their high-powered rifles in their Weasels and the newer, more powerful Sno-Cats. Gunfire resounded off the still, white mountainsides. Elk lay scattered where they fell, blood leaking from their muzzles. Butchers walked among them, cutting off the animals' legs at the knees, then daisy-chaining the carcasses to the back of the Sno-Cats, which dragged them in strings to central processing areas, where the butchers field-dressed them, hung them to freeze in the cold winter air, then loaded them into trucks.

Word got around, and a CBS television crew filmed some of the process. Senator Gale McGee, democrat of Wyoming, took notice. McGee sat on the powerful Senate Appropriations Committee, which could always get the Park Service's attention by reminding it where its money came from. He had been instrumental in stopping the Big Kill of 1961–62. In its aftermath, McGee had made sure the Park Service received money to build live traps and hire two helicopters to chase the elk into them, so they could be given away to his and other states' fish and game departments to build stocks for hunters. He was not happy that the equipment was not being used as he had intended. McGee made it clear that if he didn't get some cooperation from the Park Service, he would close the agency's regional office in Omaha, which oversaw the Rocky Mountain parks, and have its staff transferred to Wyoming, where he could keep an eye on them.

McGee subpoenaed the superintendent of Yellowstone and the director of the Park Service from Washington, and perp-walked them before an audience of his constituents at Senate subcommittee hearings in a courtroom in Casper, Wyoming. Starker Leopold, now chair of the Park Service's Advisory Committee on Natural Sciences and about to be appointed the agency's chief scientist, was flown in from California to defend the Park Service.

On the morning of March 11, before the hearings convened, McGee got Secretary Udall on the phone and secured a promise that the shooting would stop. Although a deal had already been struck, the hearings went forward for effect. During the proceedings, when the Park Service director was questioned about technical matters, he deferred to Starker Leopold. At one point, someone yelled from the back of the audience for Leopold to speak up. He leaned toward a microphone in front of him. It didn't seem to do any good. Then he realized it was a radio station's microphone, not a loudspeaker. All over Wyoming, guys with gun racks on their pickup trucks were listening to him explain why elk that otherwise might be neatly wrapped in butcher paper in their freezers were now on their way to the Blackfeet reservation in Montana.

Sitting quietly in the audience during McGee's show trial was a forty-year-old biologist from Grand Teton National Park who was about to take over wildlife management at Yellowstone. His name was Glen Cole. Watching the director of his agency being pilloried by Senator McGee, Cole made note of the political reality. Whatever he did with elk once he got there, it wouldn't involve shooting them. And that was fine. It was time nature was allowed to sort itself out on its own for a change.

16

COLE

YOU ARE SUPERVISORY research biologist for more than just Yellowstone?" Assistant US Attorney Spivak asked Glen Cole on the witness stand in Los Angeles in 1975. Cole was a balding man of medium height with big ears, glasses, and a neatly trimmed mustache over a small mouth with pursed lips.

"Well, I was supervisory research biologist of Yellowstone, Glacier, Grand Teton, Rocky Mountain—the four Rocky Mountain parks," Cole responded, "up to about 1972, when those duties were assumed—other than Yellowstone—by the regional scientists."

For Starker, the existence of such a person as Cole was a coup. There had not been one before Starker leaned on the system in the name of science. "A greatly expanded research program, oriented to management needs, must be developed within the National Park Service itself," he had written in the Leopold Report. It had been an uphill battle from the beginning. The upper echelons of the Park Service were dominated by rangers whose focus was on daily operations and the construction and maintenance of facilities. For its part, Congress saw the national parks as resorts, and looked upon ecological research as the province of the Bureau of Sport Fisheries and Wildlife. It was just ironic that the first incumbent for that chief research biologist job for the northern Rockies, Glen Cole, became a champion of a let-nature-take-its-course philosophy.

Still, Cole was far easier to deal with than the Craigheads.

ASIDE FROM THEIR shared interest in wild animals, Glen Cole could not have been more different from the dashing, daring Craighead twins. The brothers documented their exploits in books, magazine articles, television shows, and thousands of their own photographs and films. Cole, on the other hand, was a quiet, introverted man who was sufficiently uninterested in having his picture taken that his own family has almost no images of him to speak of. Like the Craigheads, he served in the Navy during World War II. But after, when the twins regaled *National Geographic* readers with their survival on Pacific islands, Cole said so little about the war that his wife and daughters didn't know the most basic facts about where he'd served and what he'd done.

Cole was an expert horseman who did not dress like a cowboy. He drove a tan Volkswagen Beetle, but unlike Zetterberg's, it was a mousy shade of beige without a convertible top. His only concession to vanity was the comb-over he maintained in the late 1960s, when everyone else had lots of hair, but his scalp had mostly given up producing it. He often went into the field wearing a khaki trench coat and a short-brimmed golf hat, looking for all the world like a suburbanite on his way to the corner store for a newspaper and a quart of milk.

After the war, Cole went to college on the GI Bill and got a master's degree in wildlife biology at Montana State, in Bozeman. His graduate advisor was Donald C. Quimby. Quimby had done his dissertation on a rodent known as the jumping mouse. He was the first wildlife scientist at Bozeman. He formed a tight relationship with the Montana Department of Fish and Wildlife, and Montana State became the vocational school for that department's wildlife biologists.

Quimby's instruction focused on the relationship of wild animals to their habitat, and required reading for his students included the classic work of Paul Errington. Errington had been a student, and later a close colleague, both of Aldo Leopold and Herbert Stoddard. However, Errington arrived at very different conclusions than Aldo about the importance of predators in regulating prey populations. Errington believed that predation was "compensatory"—that animals taken by carnivores would have died by some other means, anyway. When the range reached its carrying capacity, new animals were forced to occupy poorer-quality homes where they were more vulnerable to predation, while increasingly competing for limited food with members of their own species. So, if they were not taken by a meat eater, they would starve, freeze, or die of disease. Therefore, predators did not regulate prey, argued Errington, habitat did. This idea would have a strong impact on Cole's future work.

For his master's thesis, Cole studied herds of pronghorn that were invading farmers' alfalfa fields in eastern Montana. In 1954, he went to work for Montana Fish and Wildlife as a supervisory biologist. Cole worked for the state until 1962, when he accepted a position as a wildlife biologist at Grand Teton National Park.

With its north boundary nearly touching Yellowstone's southern one, Grand Teton National Park was created in 1929 to preserve a forty-three-mile-long wall of jagged, snowcapped peaks stretching north to south along the state line between Jackson Hole, Wyoming, and the pineries and potato fields of southeastern Idaho. Cole's government house was located at the park's headquarters in Moose, in the Jackson Hole valley, about twenty minutes' drive north of the town of Jackson. Moose wasn't really a town—just a post office, a store, park administration buildings, maintenance shops, government housing, and a little log church, the Chapel of the Transfiguration, in the sage amid clusters of quaking aspens, cottonwoods, and pines along the western bank of the Snake River.

The Chapel of the Transfiguration had been constructed around the time the park was created, with a large window over its altar framing a view of the tallest peaks of the Tetons—the so-called Cathedral Group—towering over six thousand feet above the Valley floor. As is true of *eden*, the word *cathedral* comes up over and over in connection with the national parks. Monumental scale and verticality trigger feelings of religious awe in human beings. From the Chapel of the Transfiguration or anywhere else in Jackson Hole, the Cathedral Group was like a stained-glass window. Your eyes were always drawn upward to it. At first light, when the valley was wrapped in deep indigo, the peaks would be bathed in a pink-and-orange glow. Sometimes, in the late afternoon, a summer thunderstorm would break up and the westering sun would shine under the clouds from the Idaho side, illuminating the final squalls of rain through the gaps between the peaks in brilliant yellow shafts, out over the valley. Even among scientists and people who would not consider themselves traditionally religious, living there had a way of making a person consider what was eternal, and the ephemeral quality of human dominion over the earth.

Coming to work at Grand Teton in his late thirties after a decade of shooing antelope out of alfalfa fields and ensuring a supply of venison and elk steaks to Montana hunters, Glen Cole underwent something of a transfiguration himself. "He recognized his destiny there," observed his wife, Gladys. Rachel Carson's *Silent Spring* came out the year Cole arrived, which he read, and Aldo Leopold's *A Sand County Almanac* became his favorite book. His well-thumbed copy was never far from his easy chair in the living room.

Perhaps even more influential for Cole was his association with the Muries. Three quarters of a mile down a dirt road along the river from Park Headquarters, the Murie brothers and their wives, Margaret and Louise, two half sisters from Alaska, shared a complex of log cabins that had once been a dude ranch. When they weren't at Mount McKinley, Adolph and his wife, Louise, lived in a bigger log house there, and Olaus and Margaret in a smaller one. Olaus died in October of 1963, not long after Cole came to Grand Teton, but his younger brother, Adolph, a slight man in his sixties with a great shock of gray hair whom everyone called Ade (pronounced "aid"), was by then the grand old man of wildlife biology in the national parks, and became a friend and mentor to Cole. Gladys Cole and Louise Murie were both excellent cooks, and the couples were often at each other's homes for dinner.

At the time, wildlife biologists were a rarity in the Park Service. Douglas Houston, a graduate student who came to Grand Teton to study moose under Cole in 1963, may not have been exaggerating when he said he thought that other than himself and Cole, there may not have been more than four or five in the whole country. Not much had changed since Charles Adams's complaint in 1925. In most national parks, responsibility for managing relations with nature was invested in foresters, whose education, as Adams had pointed out, prepared them for industrial forestry. At Yellowstone, with its long history of concern about overgrazing by elk, nature management was handled by a grazing specialist instead of a forester.

In the face of what Adams, George Wright, Starker Leopold, and Adolph Murie all agreed had been a long record of recklessly unscientific interventions in the ecology of national parks, Wright and Leopold espoused carefully researched restoration, and Murie took up the position that the best way to preserve parks was to make them off-limits to human control of all kinds.

"I could list many management suggestions that I personally have opposed, such as wolf control, coyote control, insect control, lodgepole blister rust control, excessive fire control, mosquito control . . . exposed garbage dumps in bear country, etc., etc.," he wrote. "This may seem like a negative attitude, but in a national park opposition to unnecessary management can more correctly be termed a positive attitude."

By the time Cole read *Silent Spring,* Murie had been objecting to the use of pesticides in national parks for over a quarter century. As early as 1937, he had issued a carefully reasoned objection to Yellowstone National Park's plan to spray firs infested with native bark beetles near Mammoth Hot Springs:

> The beetles have for some reason prospered and increased. Possibly their parasites have come to misfortune or are lagging a little behind their hosts . . . or perhaps the trees have become weakened and sap-poor through unfavorable weather conditions or old age, or both, or for some other reason . . . The story of the rise of the beetles will continue and possibly next year the parasites may catch up with the beetles and bring disaster to the prosperous

> community . . . To permit this story to progress with a natural sequence of events is, in my opinion, the particular purpose of the National Park Service. Therefore, I deem it contrary to Service policy to carry out the insect control work . . . and thereby work against nature.

Nevertheless, spraying continued for almost three decades after that as Murie continued to oppose it, along with new insults such as the opening of Yellowstone and Grand Teton national parks to snowmobile use and the widening of a road at Mt. McKinley. In 1965, the year of his retirement, the Park Service sprayed over 160,000 pounds of the pesticide ethylene dibromide mixed with more than 500,000 gallons of diesel oil on 42,000 acres of Grand Teton, where Murie lived. At this point Cole, who had succeeded Murie as Grand Teton's resident wildlife biologist, took up his mentor's noninterventionist standard in a memo to Grand Teton's superintendent, Jack Anderson. In it, he pointed to Rachel Carson's indictment of DDT and the fact that many of the trees infested with beetles were overmature and would die anyway, from one cause or another.

Adding to Cole's dismay about the artificiality of what was supposed to be wild nature, as the new wildlife biologist he became the park's representative in a byzantine system of agreements for the management of the southern Yellowstone elk between the Park Service, the Forest Service, the State of Wyoming, and the Bureau of Sport Fisheries and Wildlife, which operated the Elk Refuge. It was a wholly unnatural situation. Human interests and politics governed the southern elks' lives from birth to stewpot. They summered in the high pastures of the Grand Tetons, Yellowstone, and the Shoshone and Teton National Forests, and in the autumn they walked through rifle fire to where the village of Jackson blocked further migration. There, in a marshy meadow on the north side of town, they spent the winter shuffling around in the snow, waiting for their rations of government hay.

To make matters worse, the political logrolling required to get private ranchlands in Jackson Hole added to Grand Teton National Park in 1950 had resulted in a concession allowing the deputization of private hunters to take elk in the park, ostensibly to assist with herd reduction. However, the hunters didn't kill enough elk to make them worth the trouble of policing, and they sometimes shot animals they weren't supposed to. In 1961 alone they killed several coyotes, eleven moose, and two bears.

Under the tutelage of Ade Murie, Cole began questioning all the managerial meddling in what was supposed to be a naturalistic regime of a national park. Houston remembered Cole asking rhetorically, "How would all of this organize itself if we weren't cropping game here?" Another Park Service coworker remembered Cole posing the question, "How did all these things get by before we came along?" Cole's resistance to spraying and to wildlife cropping was clearly in line with the Leopold Report's interest in avoiding artificiality. But in the rest of his sentiments Cole was far less in line with Starker than with his friend and mentor Adolph Murie.

All of this was on Cole's mind when he got the promotion to chief research biologist of the northern Rocky Mountain parks and watched Park Service director Hartzog, Yellowstone superintendent McLaughlin, and Starker Leopold get crucified in McGee's hearing in March of 1967.

Over the Memorial Day weekend in 1967, Cole packed up his family and drove them to Yellowstone to start his new job. Down the hall from his new office, on the third floor of the old stone cavalry barracks, was the office of John Good, the park's chief naturalist. Good had worked as a petroleum geologist before he came to the Park Service, and Cole evidently thought he knew a little paleoecology. Cole was thinking about whether, left to their own devices, elk would find a balance with their food supply. One day he appeared in Good's office doorway.

"John," he asked the chief naturalist, "can you think of anything in the paleontological record that would indicate that a species has ever destroyed its own food supply, thereby destroying itself?"

Good thought for a moment. "No," he answered.

"Hmm," said Cole thoughtfully, then turned and walked out.

JUST BEFORE COLE'S arrival that spring, the Craighead brothers delivered to Yellowstone superintendent McLaughlin a 113-page report called for by their 1959 contract with the Park Service, containing their recommendations on improving management of Yellowstone's grizzlies. Seven years of radio tracking the far-ranging bears led the brothers to recommend that grizzlies be managed cooperatively between the park, surrounding national forests, the Bureau of Sport Fisheries and Wildlife, and neighboring states. The document acknowledged the unnaturalness of the dumps but strongly recommended against closing them rapidly. In the event a decision was made to close them, the brothers suggested that elk carcasses be helicoptered into backcountry "bait stations" to draw grizzlies away from populated areas during the transition. These areas and others where grizzlies congregated should be closed to visitors, to separate bears and people. The Craigheads further recommended eliminating every possible alternative source of human food, with rigorous policing of the campgrounds, secure trash facilities, and visitor education. If shutting down the dumps was not handled with great care, they wrote, "The net result could be tragic personal injury, costly damages, and a drastic reduction in the number of grizzlies."

The way the Craigheads told it, the trouble between them and the Park Service began the fall of 1967 with the transfer of the new superintendent, Jack Anderson to Yellowstone from Grand Teton. But tension had already developed with the outgoing superintendent, John McLaughlin. To enhance the funding the grizzly study was getting from the National Science Foundation, the National Geographic Society, the Atomic Energy Commission, the Philco Corporation, and the Park Service, in 1962 Frank Craighead had gone to Montana senator Lee Metcalf seeking additional government funds. He was successful, but when the additional $6,000 came through, it was routed through the Park Service budget beyond McLaughlin's discretion about how to spend it. There was a perception that the Craigheads had gone over the superintendent's head.

Then, in June of 1967, the Hollywood studio producing the National Geographic television special on the brothers' grizzly work mailed a copy of the final script to McLaughlin. McLaughlin was not happy. Produced for commercial broadcast, the script didn't resemble those sonorous short documentaries shown in park visitor centers. McLaughlin thought it was entirely too much about the Craigheads' exploits and not enough about the Park Service. It didn't contain enough hard scientific information, he complained, and there was too much emphasis on the handling and manipulation of bears and not enough on Yellowstone as a natural place with free-roaming wildlife. Overall, there was a sense that the two larger-than-life celebrity biologists were a little too large for Yellowstone. This tension arose at the worst possible time: on the eve of the arrival of a new Yellowstone superintendent and the most shocking grizzly attack in Park Service history.

17

THE NIGHT OF THE GRIZZLIES

IN 1967, TWO of the most intractable problems with managing nature in the national parks blew up within twenty-four hours of each other at Glacier National Park. At Yellowstone, the events of that summer precipitated a series of actions by the Craigheads and the Park Service that eventually led to the termination of the brothers' research and the filing of the Walker lawsuit in Los Angeles.

The summer of 1967 would be the driest in the fifty-three years in which records had been kept at the West Glacier ranger station. The blueberry crop failed, putting pressure on bears to find other food. Crews fanned out in the park, as they had for years, trying to control white pine blister rust—the exotic Asian fungus that killed the park's beautiful, high-altitude whitebark pines—by uprooting the fungus's intermediate hosts, wild currant bushes. Lloyd and Grace Parratt had stayed in California to watch over Smitty, but one of their older sons, Monty, was working at Glacier on a blister rust crew. Monty had married, and his young wife, Laurel, was employed at one of the park hotels.

In mid-June, a skinny, long-faced grizzly with a patchy, mangy-looking coat had been seen turning over unsecured trash cans at a cluster of private cabins at the northeast end of Lake McDonald, in the southwest part of the park. The bear was brazenly unafraid of summer residents who tried to scare it off. Half a mile west of the cabins along the lakeshore was the trail into the next drainage to the north. The trail gained over 2,000 feet of elevation in a little over two miles to the crest of Howe Ridge, then dropped about 1,300 feet through the timber and blueberries to Trout Lake, a mile-and-a-half long, deep-sapphire pool in the valley of Camas Creek. In the winter, the mountainsides on either side of the drainage were raked by avalanches, which had carried hundreds of trees into the lake. Weathered silver-gray by the elements, the trees lay in a huge log jam at the outlet, where the trail reached the lake. Next to the log jam, there was a backpackers' camp. As was typical at the time, the Park Service had installed a regulation steel campfire grate, but no facility for storing food safely away from animals.

In July, there were multiple reports of a bear that matched the description of the mangy, long-faced animal from the other side of the ridge at Lake McDonald, invading camps at Trout Lake, where it tore up camping equipment, stole food, and terrified campers. The empty, smashed food containers from its depredations added to a quantity of other trash left by inexperienced and poorly trained backpackers, accumulating in the area of the campsite. In the first week of August, a group of Girl Scouts and their leaders suffered two days of escalating skirmishes with the animal there before fleeing with what remained of their mangled equipment. As they were leaving, one of them took a photo of the grizzly walking out on the logjam, which appeared in the local newspaper. On Saturday, August 11, a man and his son were charged by an aggressive grizzly at Trout Lake and took refuge in trees for two hours as the bear ripped up their backpacks and consumed their food. As was true of the handling of the attack that preceded Smitty Parratt's mauling, even after the repeated incidents at Trout Lake, the trail to the lake wasn't closed to the public. To be fair, other problems were stretching the Park Service's resources at that moment.

By the end of July, with the drought, there had been hundreds of lightning- and human-caused fires in the northern Rockies. By early August, temperatures at Glacier National Park's lower elevations were in the mid-nineties and humidity was as low as 10 percent. On the evening of August 11, a dry lightning storm moved in. Perched on their glass-insulated stools in their towers, fire lookouts recorded more than a hundred ground strikes in Glacier National Park alone. No measurable rain fell. As the storm passed, wisps of smoke rose from the forests, and in the morning the lookouts called in twenty-three new fires in the park. There were more in the neighboring national forests.

In addition to seasonal rangers, such as Lloyd Parratt, and the fire lookouts, firefighters, maintenance crews, and other Park Service workers, the populations of American national parks swell during each summer with an even greater number of temporary staff for hotels, restaurants, stores, and gas stations. Today, many of these people are immigrant workers or retirees, but in those days they were mostly students. They came to save a little money for college, to meet other young people, and to spend their days hiking, backpacking, or climbing mountains. In 1967 there were more than eight hundred of them working for various concessions at Glacier National Park.

Julie Helgeson had just finished her freshman year at the University of Minnesota and was employed in the laundry at East Glacier Lodge. She was a friendly and outgoing young woman of nineteen with an unmistakable Upper Midwest accent. Her family was descended from Scandinavian immigrants and she looked it. She'd been runner-up for Miss Albert Lea, Minnesota. At East Glacier she befriended Roy Ducat. He was about a year younger and was about to enter his sophomore year at Bowling Green State University in Ohio. The two of them had a couple of days off and decided to hike to Granite Park Chalet.

Granite Park Chalet was one of Glacier's European-style mountain huts, a sturdy, two-story structure constructed of native stone and wood, which stood in the rocky alpine meadows at the end of a prominence projecting west from the Continental Divide. A loop of trail descended from the chalet to an informal backpacker camp a third of a mile north and downhill, at the upper edges of a forest of subalpine fir. A couple of hundred feet up the trail beyond that was a log cabin used by trail crews, and from there the far end of the trail climbed a short distance back up to the chalet. The chalet was operated by seasonal employees of the same company that ran Julie's hotel. For years, they'd been dumping kitchen waste and garbage into a swale near the trail down the hill to the campground, about two hundred feet from, and within view of, the chalet. Watching the grizzlies feeding on garbage was a well-known attraction of a stay at the chalet. After dark on summer evenings, the staff took the buckets of slop out, then made an announcement, and the guests trooped out to the lodge's porch. The bears arrived, walking up the trail from the direction of the campground, and the guests watched them feed, spotlighted by the staff's flashlights. The practice went on for years as hundreds of people came and went. In 1962, a man was clawed in his sleeping bag at the backpacker camp but not seriously hurt. In 1965, while researching the Parratt case, Stephen Zetterberg hiked into the chalet overnight and witnessed the feeding. The following August, a seasonal ranger sleeping in the backpacker site was awakened by a grizzly pulling on his sleeping bag. Unhurt, he climbed a tree and stayed in it for the rest of the night.

The Park Service knew about the feedings. A small incinerator was provided for the chalet, and at the end of July in 1967, Glacier National Park's chief biologist asked a seasonal ranger to check on the situation. The ranger hiked into the chalet three times. On July 31 he reported that the chalet staff had told him that the incinerator was not big enough to handle their output, and they were still feeding the bears. On his second visit, on August 1, he reported that he watched the bears being fed. He made a third trip to watch the bears with another ranger and his wife and the curator of the museum at park headquarters, after which he reported that there were from six to eight grizzlies at Granite Park. Nothing was done.

On August 12, 1967, Julie Helgeson and Roy Ducat hitchhiked to a trailhead on Logan Pass and from there hiked seven miles along the west side of the Continental Divide to Granite Park. Arriving in the early evening, they left their gear at the backpacker site and walked up the loop trail to spend the evening at the chalet. Close to sunset, around 8:00 p.m., they walked back down to the campsite and spread out their sleeping bags in a patch of meadow between the trees, surrounded by lush towers of false Solomon's seal, bear grass, and glacier lilies. It was a warm evening, the sky was slightly hazy with smoke from the fires, and as dusk came there was a first-quarter moon. They talked for a while, then fell asleep as the summer constellations of Cygnus, the swan, and Aquila, the eagle, wheeled dreamily through the night sky against the inky points of the conifers.

ELEVEN MILES SOUTHWEST of Granite Park Chalet, Michele Koons, a nineteen-year-old college student from San Diego, was employed for the summer at the gift shop at Lake McDonald Lodge. Known to her friends as Micki, she was a tiny, pretty girl who meant to make her mark in the world. In high school she was president of her senior class and was voted "most likely to succeed" and "outstanding senior girl" by her classmates. She played Mary in her church's annual Christmas pageant.

On the same morning Julie Helgeson and Roy Ducat left on their hike to Granite Park, Michele and four fellow seasonal workers left Lake McDonald on a weekend backpacking trip on the West Lakes Trail over Howe Ridge, to Trout Lake. On their arrival there, they encountered the man and his son who had been treed by the grizzly, who were on their way out. It was too late to go back, so Koons and her party set up camp at the backpacker's site. They caught some fish, and in the early evening they were cooking the fish and some hot dogs over a fire when a mangy-looking grizzly entered their camp. Living up to its reputation, the bear had no apparent fear of human beings. It chased the four young people away, rooted around in their gear, and consumed much of their food. They abandoned their site and established a new camp nearby, where they built a large fire and eventually went to sleep with their sleeping bags laid out in a circle close around it.

AT GRANITE PARK, around 12:45 a.m., Roy Ducat awoke to the sound of Julie Helgeson whispering earnestly for him to lie still and play dead. A moment later, a tremendous blow sent Ducat flying through the air. He felt a pair of jaws close on his shoulder and teeth grated on his bones.

He remembered the same advice Smitty Parratt and his party had been given, to play dead. It took tremendous willpower not to cry out. The attack ceased, and Ducat opened his eyes to see the dark bulk of the grizzly standing over Helgeson. The bear seized the young woman, and he heard her scream, "It hurts!" Then he heard what he described to investigators as bones crunching. Julie kept screaming, and her cries were heard by people a third of a mile away at the Granite Park Chalet, as well as by others camped nearby at the Park Service trail crew cabin. One witness heard Julie cry out for her mother. Then Roy heard Julie's cries receding, and he realized that she was being dragged downhill, away from their camp.

Bleeding, and with one damaged arm dangling uselessly by his side, Roy got up and staggered to the trail crew cabin, where he collapsed in front of the backpackers who had been awaked by the screams. One of the campers ran up the hill to summon help, and a group of volunteers hiked down and carried Ducat back up to the chalet, where an impromptu team of three doctors, who had arrived separately to stay the night, administered first aid. The commotion awakened the lodge's guests, including Monty Parratt's wife, Laurel, who had hiked in to stay at the chalet with friends.

A twenty-two-year-old botanist in her first summer with the Park Service as a ranger naturalist had led a group of park visitors on an overnight hike to the chalet, and that evening, with the others, she had watched the staff feed the local grizzlies. Sometime after midnight she was awakened by a commotion inside the chalet. When she heard what had happened, she activated the lodge's two-way radio and tried to figure out how to operate it. Her urgent calls for help were received by a ranger in a patrol car who relayed them to the park's fire dispatcher. Every able-bodied employee the park could muster had been sent out to battle the fires, among them Monty Parratt and his blister rust crew, who were fighting a fire on Apgar Mountain. Monty was carrying a fitful World War II army-surplus backpack radio equipped with vacuum tubes and weighing forty-six pounds. Over this device, garbled and mixed with fire traffic, he heard snatches of transmissions about a bear attack at Granite Park Chalet. He knew his young wife was up there, and after the Parratts' trauma with Smitty, for the rest of the night and into the next morning Monty waited in agony for news of Laurel. Meanwhile, up at the chalet, Laurel worried about Monty, out in the dark somewhere, among the flames.

Summoned by the fire dispatcher, a twenty-eight-year-old pilot who had flown Army helicopters in Vietnam undertook a perilous night mission in the park's contract firefighting helicopter to bring an armed ranger and medical supplies to Granite Park and extract Roy Ducat. At the chalet, volunteers made bonfires to guide him in and tried to light the tiny landing zone between the chalet and an accessory building with flashlights. As the pilot was hovering slowly down on his final approach, someone realized there was a steel flagpole there, and people began urgently waving flashlights to warn him off. The pilot was blinded by the lights reflecting on the helicopter's canopy and pulled out. A radio conversation ensued, the lights were dimmed, and he landed safely.

Once on the ground, the ranger with the rifle led a party of volunteers down the trail to the backpacker camp, lugging a bonfire in a large washtub they thought would scare away bears. They located Roy and Julie's disheveled campsite, and followed a blood trail down a sloping meadow until it became indistinct. Searching farther down the hill, they heard moans. The bear had dragged Julie 342 feet from her campsite. She was conscious, but the grizzly had inflicted massive injuries to her legs, the flesh of one of her forearms was eaten off, leaving bare bones, and she had lost a lot of blood. Worse yet, she was suffering from the same condition that might have killed Smitty Parratt. A gurgling hole through her chest and the bulging accessory muscles in her neck signaled that enough air had entered outside her lungs to put her in grave danger of suffocation. She was carried up the hill to the chalet.

In the chalet's dining room, the young naturalist and the three doctors had set up an impromptu operating theater on two dining tables. Led by an experienced surgeon, the team worked feverishly to save the young woman. Another overnight guest, a Jesuit priest, bent over her face and spoke to her soothingly as the doctors struggled against the seriousness of her injuries. When it became apparent that Julie Helgeson was dying, the priest baptized her with water from the kitchen and administered last rites. Still conscious, she responded weakly as the priest finished the ceremony. Then, witnesses told a reporter, she gasped, hiccuped, and died.

NINE AIR MILES away, Michele Koons and her party were slumbering in their sleeping bags on the shore of Trout Lake. Their fire had burned low and a bag of cookies salvaged from the bear's rampage at their first campsite sat on a log nearby.

Sometime that night, members of the group woke to the sound of splashing from the lakeshore. A grizzly entered the camp, took the bag of cookies, and left. A few minutes later it was back, sniffing around their sleeping bags. One of Michele's male companions bolted for a tree and climbed it. The others jumped out of their sleeping bags and headed down the lakeshore, where they, too, established themselves in trees. The first boy was up a tree directly above the campsite, and from there he saw the bear approach Michele Koons, who was still in her sleeping bag. He yelled for her to get out and run. Panicked, she called back that she couldn't; the bear had hold of the zipper. Then she screamed, "He's ripping my arm!" And then, "He's got my arm. . . . My arm's gone! Oh, my God, I'm dead!" Then she was silent, and the boy saw the dark shape of the bear dragging her, still in her sleeping bag, up into the forest and heard what he later described as bones being snapped.

The rest of the party remained perched in the trees until daylight, then climbed down and hiked out to get help. Rangers hiked in with rifles to hunt the grizzly and to look for Koons. They found her sleeping bag twenty feet up into the forest and followed a trail of blood and clothing another 87 feet to her remains. So much of her had been eaten that it was no longer possible to tell the body was female.

THE NEWS OF the first two fatal bear attacks in Glacier National Park's history on the same night hit Montana newspapers on August 14 and was picked up by the wire services and national television. The question everyone was asking was "why?" One story suggested that bears might be attracted to sleeping bags, since both women were in them at the time of the attacks. Others suggested that the violent lightning storm the night before the attacks might have agitated the bears.

When reporters asked Glacier National Park superintendent Keith Neilson what might have provoked the grizzlies, he acknowledged the failure of the park's blueberry bushes to set fruit but reasoned that the bears had a lot of other things to eat. He allowed that there might be some truth to the lightning theory but mentioned nothing about the habituation of grizzlies to human food and garbage. When John and Frank Craighead were asked to comment, they discussed grizzly behavior in general, but Frank said that it would be irresponsible to conjecture about causes before a thorough investigation. The Park Service had already dispatched Chief Research Biologist Glen Cole to oversee one.

Two days later, the *Billings Gazette* reported the account of an off-duty Montana state highway department public-information officer who had hiked to Granite Park with his wife and child to stay over the night of the attacks. That evening, he witnessed the chalet's staff announcing what was apparently a nightly entertainment for guests, and he saw them set out the kitchen scraps within view of the building. Presently, bears appeared out of the darkness, walking up the trail Helgeson and Ducat used to get down to the backpacker campground. The *Gazette* further revealed that a park naturalist had reported the practice to superiors about two weeks before the attack, and nothing had been done about it.

A week after the two deaths, the *Seattle Post-Intelligencer* reported that Superintendent Neilson had admitted to a connection between the attacks and the bears' acquisition of human food and garbage but placed the blame on chalet workers and backpackers like the Koons party, who were careless about storing food and disposing of their garbage. The closest he came to accepting responsibility was to say that the Park Service needed to crack down on visitors who fed bears.

In Missoula, John Craighead saw an opportunity to elevate into a public discussion the actual issues behind the attacks. John McLaughlin, to whom he and Frank had presented their management recommendations that spring, was the second park superintendent they'd dealt with in eight years, and now he was leaving without responding to their report, to be replaced by a third. In an audacious move, John Craighead released the report to the newspapers. It was, after all, a public document. But it surfaced as the National Park Service was nervously dissembling about the causes of the Glacier attacks, and it set up the Craigheads as independent critics of the Park Service.

CRAIGHEADS HIT GRIZZLY POLICY, read a headline that weekend in the *Livingston Enterprise,* the local newspaper for residents of the Park Service complex at Mammoth Hot Springs. McLaughlin was even more unhappy, but he was leaving. Jack Anderson wasn't scheduled to replace McLaughlin until October, but his mandate already seemed clear. He needed to get some control over the elk and grizzly issues. And he definitely needed to get a handle on the Craigheads.

18

NATURAL CONTROL

AFTER LEAVING YELLOWSTONE in September, Superintendent John McLaughlin took up his new post at Sequoia and Kings Canyon National Parks' headquarters in the oak-studded foothills at Ash Mountain. In October, he drove to Berkeley for a meeting called by Starker Leopold—who had been named the Park Service's chief scientist—with fire experts at the Forest Service's Berkeley research shop, the Pacific Southwest Experiment Station. Leopold wanted the Forest Service's input on carrying out the first real prescribed burn at Sequoia—not just burning brush piles, or something like that damp, diesel-soaked affair at Yellowstone—but actually setting fire, under scientific specifications, to the woods. Bruce Kilgore, a tall, deep-voiced Park Service biologist and former editor of the *Sierra Club Bulletin,* also attended. An acolyte of Leopold's since his undergraduate days at Berkeley, Kilgore remembered going on a field trip with the professor in the coastal hills north of San Francisco. Looking out over the thick brush, Starker had told the students: "Someday the government will be burning all of this." Now, Kilgore would be the point of the spear, implementing the Leopold Report at Sequoia and Kings Canyon.

However, there was considerable skepticism from the Forest Service experts at the meeting in Berkeley. Starker listened for a while, then interrupted the meeting to quietly, but emphatically, put the matter to rest.

"We came to this meeting to get ideas on where and how to go. We are not asking your opinion on *whether* we should go. We want to know what the best program is. In fact, we are *going to prescribe-burn*!"

AT MAMMOTH HOT Springs over Memorial Day weekend in 1967, Glen Cole, his wife, and their two little girls moved into one of a cluster of modest ranch-style homes, like those on military bases, that sat on a dry hillside below park headquarters amid a few mountain junipers that had been high-lined by hungry elk since at least the early 1930s.

Cole worked, and when he wasn't working, he worked at home. On weekends, his idea of a family outing was to drive over to the Lamar Valley and watch the elk. Or they'd drive over to Bozeman, where Cole would talk shop with his friend Dick Mackie, a professor of wildlife management at Montana State, and Gladys and the girls would go shopping. When he was driving Glen didn't say much. In the evenings and on weekends, he'd sit in an easy chair in the living room, writing in pencil on a pad of lined government paper, which he'd give to his secretary to type up. Sometimes he would smoke his pipe, and the redolent tobacco smoke would waft through the house.

Cole was writing about elk. For decades, rangers and scientists had seen the threadbare condition of the Northern Range as evidence that there were too many of them. But could it be that the belief that the "natural" condition of Yellowstone was lush aspen forests and lightly grazed meadows and sagebrush steppes was the same kind of arbitrary aesthetic judgment that drove spraying for spruce budworm and pine beetle? A belief that things ought to look pretty? Could what had been seen as the "overgrazed" condition of the Northern Range in fact be the mechanism by which nature controlled the elk population—that is, by starvation?

Were there really too many elk?

"I personally have some very strong feelings against prevalent attitudes that we have excessive or surplus members of ungulates throughout the park," wrote Cole in a letter to incoming superintendent Jack Anderson. "Over large areas we are extremely short of native ungulates if we consider that the Park's primary purpose is to provide visitors with the opportunity to see native wildlife scenes and maintain some semblance of a natural ecosystem."

That autumn, under Cole, Yellowstone National Park set forth a new policy for managing elk called "natural control." Among the educational purposes of the park—wrote the unsigned author of a briefing paper, believed to have been Cole himself—was to encourage an understanding of the elk's role in maintaining a "balance of nature." Visitors ought to be able to see wildlife as it had occurred in primitive America, the paper said, echoing Starker Leopold.

But the policy differed from Leopold in how to accomplish this. In order for visitors to see a natural ecosystem, according to the policy, natural regulatory processes, not human controls, had to be allowed to govern the prevalence of animals. Elk had lived in Yellowstone for thousands of years. They could not have persisted had natural controls not existed to keep them in balance with their plant food sources. Another unsigned memorandum, issued under Cole that December and believed to have been written by him, stated: "The most favored form of management is to rely upon natural controls to regulate animal numbers whenever possible. This automatically results in a balance of nature."

Under this "balance of nature," winter weather and starvation would do part of the job of regulating elk herds. But Cole also referred mysteriously in his writings to the persistence of scattered wolves in remote areas of the park. The numbers of "probable sightings" of wolves rose rapidly from the time of Cole's arrival at Yellowstone to peak at forty a year at the end of the 1960s. There is no convincing evidence, says Yellowstone's later wolf biologist, who has studied the animal for well over three decades, that these late-1960s wolves actually existed.

What is apparent is that Cole, under the tutelage of Adolph Murie, had become leery of so much ill-fated messing around with nature. Cole wanted to believe that the balance of nature was intact at Yellowstone, and he believed that you couldn't see it work until you stopped constantly doing things to it. "Glen was thinking for an ideal world, and it wasn't an ideal world," observed his friend Dick Mackie.

19

BAD BLOOD

When Harry Walker turned twenty-one in 1968, the world was rocked by change. Yet the Walkers changed little. In the evening Wallace would read the *Anniston Star* and watch the news in the living room. The assassinations of Robert Kennedy and Martin Luther King Jr., the Tet Offensive in Vietnam, the student takeover at Columbia University, bombs going off, and rioting in Chicago, New York, Boston, Detroit, Kansas City, Newark, Baltimore, and Washington, DC, passed before him on the flickering screen, but politics weren't discussed at the Walker dinner table. The Walkers were basically apolitical, but they were not amoral. Wallace and Louise operated by their own high ethical standards. Louise gave her fellow farm wives driving lessons to increase their independence from their husbands, and her children grew up playing with the kids of African American neighbors.

Harry kept his hair short, attended his National Guard trainings, and spent his spare time hunting deer and opossum and angling for mud cats and bluegill in Choccolocco Creek, baiting his hook with catalpa worms collected from a tree that shaded the farmhouse. He doted on his dogs—a beagle he named Sloopy, after the popular song "Hang on Sloopy," and two redbone hunting hounds, Duke and Duchess. He plowed a neighbor's fields in trade for a dappled brown-and-white colt he named Comanche, and he trained the horse himself. In a home movie filmed during the 1960s, Harry stands next to Comanche, barefoot in shorts and a T-shirt, talking softly into the horse's ear. Then he encircles the animal's neck in his arms, kicks a long leg over its withers, and clambers up, over, and down the other side as the horse stands serenely still. In another film clip, Harry gallops up from the lower pasture, the young man and horse one fluid being. Harry used Comanche to herd cows, riding bareback without a bridle or hackamore, using only his voice, hands, and knees to direct the animal.

From the back steps of the farmhouse where he had performed his morning inspections as a small boy, Harry could still gaze out on the same lower pasture as a young man: sheeted in blue flowers in springtime, with fireflies rising like sparks from it on summer evenings, and yellow with black-eyed Susans in the late summer and fall. In the middle of the field stood a single old loblolly pine. Hawks, vultures, and flocks of crows would settle on it. During violent thunderstorms the cows would take shelter under it, where a single lightning strike could wipe out the entire herd. One day Wallace tried to push the tree over with a bulldozer, but it wouldn't budge, so he left it, and the old pine stood upright in that pasture longer than Wallace himself.

IN EARLY 1968, after Glen Cole had decided to let elk regulate their own population rather than kill them, he and Jack Anderson turned their attention to their other major issue with park wildlife: ensuring that when the Centennial came in 1972, Yellowstone visitors wouldn't encounter beggar black bears and killer grizzlies. To deal with the black bears, Anderson and Cole continued installation of bear-proof trash cans, and Anderson ordered rangers to issue citations to visitors for feeding them, an offense that had never resulted in much more than a warning. They picked up the pace of trapping, translocating, and, if need be, killing black bears that were hanging around roads and developed areas.

In the case of the grizzlies, Cole was convinced that the dumps were the problem. They were clearly out of step with the Leopold Report's charge to eliminate "observable artificiality in any form." Feeding wild animals was an embarrassing remnant of the old days, and privately, Cole knew it had led to the death at Granite Park. He was so against feeding wild animals, he didn't even have a bird feeder at his home.

The Craigheads, however, saw the dumps differently, as a source of food high in protein and fat early in the season, when the sows emerged from hibernation, nursing their cubs. To them, Trout Creek wasn't pretty or natural, but it was a magnet drawing a threatened population to the geographical center of a landscape in which they were nominally protected, away from the sheep allotments of the Targhee National Forest, the vacation cabins of Cooke City, and all the other places where bears might be shot once they set foot across the boundary.

"John and Frank saw the Trout Creek dump as just another blueberry patch or one of those Alaskan streams full of salmon," Maurice Hornocker explained later. "Grizzlies had always gathered during certain feeding opportunities, and in a way, with regard to making sure the species survived the next few years south of Canada, this was no different."

While Glen Cole saw the anthropogenic nature of what the bears found at the dump as directly related to their loss of fear of humans, the Craigheads argued that past a locked gate up a lonely road in the hills of Hayden Valley, the associative link between food of human origin and humans themselves was missing. Bears had no idea where peanut butter came from, unless it was sitting on a picnic table in a campground full of people.

But Anderson and Cole wanted to be able to say that bears were living on natural foods by the time of the Yellowstone Centennial. As a first step toward dump closure, they decided to do something that had actually been suggested by the Craigheads. The brothers recommended that high-food-value kitchen wastes from hotels and restaurants be separated from burnable refuse such as the contents of trash barrels, and that the food be set out for bears at the dumps, and the burnables incinerated. In 1968, Jack Anderson ordered this separation for garbage going into the Trout Creek dump.

However, what the bears had been eating before wasn't just kitchen waste. There was food in the burnable waste: a half-eaten hamburger and fries wrapped up in a takeout bag, peaches with brown spots, a partial can of beans, spoiled bacon and curdled milk that had gone bad in some camper's ice chest. Burning this rather than putting it in the dump cut down the bears' food supply. What was more, it took a while to set up the separation program, and for a while, at the beginning of the 1968 season, deliveries of bear food almost entirely stopped. Then, when the food did start showing up, it wasn't enough. The reduction in nutrition was felt almost immediately in the form of increased bear activity in nearby areas of the park.

CALLED AS A witness for the defense on January 16, 1975, Jack Anderson testified to what happened. "Mr. Anderson," said William Spivak, "about this separation program in 1968 and in 1969. Were you and your staff monitoring this program and observing it?"

"Yes, sir," replied the superintendent, a big, chesty man with a chiseled face and bushy eyebrows in a full-dress green uniform with a gold badge.

"And, in your judgment, was the program a success?" asked Spivak.

"No, sir," replied Anderson. "It was not, because we had an escalation of movement of bears into three of the campgrounds that were in fairly close proximity to Trout Creek. . . ."

The campgrounds the superintendent was referring to, which became ground zero for the effect of trash reduction at Trout Creek, were Fishing Bridge, seven miles south of the dump on the north shore of Yellowstone Lake, and Canyon, eight miles north of it on the rim of the Grand Canyon of the Yellowstone. The Park Service had not finished installation of bear-resistant trash barrels at either of them when the inflow of trash was reduced at Trout Creek. Rangers worked day and night to get visitors to put their food away in their vehicles, but taking the ketchup, mustard, the jar of pickles off of the red-checked oilcloth tablecloth when people went to bed was fundamentally un-American behavior. Another area affected by the change was Old Faithful, where a local dump at Rabbit Creek, four miles north of the village, attracted bears as food was reduced at Trout Creek. In June 1968 a sixteen-year-old male grizzly tried to break into an occupied travel trailer at Fishing Bridge and was shot. About a week later, an eleven-year-old male grizzly found an ice chest left out by a camper at Old Faithful and began feeding from it. A ranger responded and darted the bear. The bear charged, and the ranger shot it.

IN THE SEVEN or so months since Jack Anderson had taken over as superintendent, feelings of suspicion and frustration had continued to grow between the Park Service and the Craigheads. In a meeting at Anderson's office at Mammoth Hot Springs on the morning of June 28, 1969, the superintendent began putting on the squeeze. He told John Craighead that Park Service director George Hartzog wanted the old employee mess hall at Canyon that the Craigheads were using as a laboratory to be torn down to beautify the park for the Centennial. Craighead asked Anderson to delay demolition, and in the event that was not possible, he asked the superintendent for alternative quarters. Anderson said he wasn't sure the park could house the study anymore.

The two men discussed the progress of trash separation at the Trout Creek dump. Anderson said he thought there were about sixty garbage cans of bear food going into the dump each day. Craighead said the deliveries seemed to stop entirely for a period at the beginning of the season, and they now looked like considerably less than the superintendent estimated. If they didn't increase, Craighead worried that bears dispersing into the campgrounds would make it a very rough summer for everybody. The conversation ended tensely.

Craighead then went upstairs to the biologists' offices on the third floor to see Glen Cole. At Cole's office there was a discussion of the northern elk herd and the Craigheads' elk research project. Cole told Craighead that the park's management of nature was about to have a new look. "They were going to manage on the basis of the ecosystem and there would actually be very little management," wrote Craighead in his notes on the meeting. "They planned to continue research, but . . . all the marked animals would have to have their markers removed by 1972." In one day, the twins had been told that they were losing their housing in the park along with the principal method by which their research was accomplished.

Meanwhile, the effects of trash separation continued to ripple into the campgrounds. On July 16, the Craigheads' research bear number 239, an eight-year-old male, was shot by an armed park visitor for trying to steal his groceries on Pelican Creek, near Fishing Bridge. Three days later rangers were trying to move a grizzly that had been hanging around Grant Village, on the western shore of Yellowstone Lake. The tranquilizer dart hit the bear in the spine, paralyzing its hindquarters, and the rangers had to put it out of its misery. On July 21 in West Yellowstone, a seven-year-old male grizzly terrorized a campground, ripping through tent walls and biting a boy on the arm. This bear too was shot. Two days later, back at Fishing Bridge, a grizzly that had for a decade been one of the Craigheads' research subjects followed its sensitive nose to some fish entrails left on a beach by a fisherman, consumed them, and bedded down for a nap in a willow thicket. Another fisherman stumbled on the bear there, and the grizzly warned him off but did not hurt him. A ranger was called to the scene, the grizzly charged him, and he shot it. On August 4 another fisherman ran into another grizzly in its daybed in the willows at Fishing Bridge, a fifteen-year-old male who'd been in the Craigheads' research population for eight years. In making its escape, the grizzly scratched the man. A ranger came out, found the bear, and shot it.

ON AUGUST 13, 1968, the first anniversary of the 1967 attacks, Glacier National Park released its official findings in the deaths of Julie Helgeson and Michele Koons. The document did not identify the feedings at Granite Park Chalet as a proximal cause of Julie Helgeson's death, only as a factor in the unusual concentration of bears to the area. To the extent that the attraction of bears to Granite Park was a contributing factor, the investigation placed responsibility for it on the chalet staff, skirting the Park Service's role in allowing the practice to continue for years, or even decades. The investigators concluded that an older female grizzly with worn teeth had killed Helgeson. The animal had been seen acting aggressively toward other bears at evening feedings, and when rangers shot her and two other grizzlies in the area of the chalet after Julie's death, Glen Cole and Glacier's wildlife technician examined the carcass and found a bad cut on one of her paws. The report suggested that pain from that injury and competition with other bears for food had agitated her enough to kill.

With regard to the fatality at Trout Lake, the report pointed to four gunnysacks of trash rangers removed from around the backpacker campsite after the incident, implying that hikers' failure to carry out their garbage and store food out of reach of bears had been responsible for the killer bear's repeated visits and, eventually, for Koons's death. The possibility that Trout Lake could have been closed to visitors after early reports of the bear's dangerous behavior there—as Superintendent Hummel admitted he ought to have done on the Roes Creek Trail in 1960—was not discussed.

What was discussed was menstruation.

In October of 1967, months before the official findings were released, Glacier superintendent Neilson had told the *Great Falls Tribune* that his experts were looking into the possibility that both victims were menstruating. He said the park had received about 150 letters theorizing on causes for the deaths, of which about 20 percent pointed to menstruation as an attractant to bears. Neilson characterized these letters as coming from people who knew about animals: circus trainers, zoo workers, ranchers, and hunters.

The August 1968 report concluded that, indeed, Michele Koons had been menstruating. Menstrual pads had been found in her personal effects. And it stated that Julie Helgeson, whose pack contained menstrual supplies, might also have been expecting the onset of menstruation. The document postulated that menstrual odor, and sweet-swelling cosmetics in the young women's possession, may have attracted the bears that killed them.*

Following the Glacier report's release the menstruation story took on a life of its own, as it was recited by rangers throughout the national parks. In some cases female Park Service employees were forbidden to work in areas where they might encounter bears during that time of the month. Later, an exhaustive study of bear attacks by Stephen Herrero of the University of British Columbia failed to find any correlation between menstruation and bear incidents. Of the various categories distinguished by Herrero, the type of attack suffered by Smitty Parratt—a sow defending cubs—will always be a risk where bears and people mix. However, in 1967, the most common cause was "food conditioning," in which bears were repeatedly rewarded with food or garbage for losing their wariness around human beings. This could be an underlying cause even in the sow-cub cases, if it led the females to bring their cubs close to people. Bears receiving food rewards in the presence of people, not menstruation, was the problem.

GLEN COLE KNEW that the Craigheads would not agree with the finding of the Glacier report and that they had a history of taking their case directly to the media. So he decided to make a preemptive strike before the report came out. On the witness stand in Los Angeles, Frank Craighead described what had happened.**

On July 9, 1968, said Frank, Cole came to see him at the Green Lab

---

* It is worth mentioning that the Park Service's report on the grizzly attacks of August 13, 1967, was written by men, who may not have known that many women have menstrual supplies in their personal effects whether they are menstruating or not.

** Frank Craighead's account of this exchange was confirmed in a signed statement by a graduate student, Jay Sumner, who witnessed it. Frank told the story in his testimony in Los Angeles in 1975, and Glen Cole, who was also a witness at the trial, never denied that it happened.

at Canyon. In the conversation that followed, Cole complained that the Craigheads were embarrassing the Park Service in statements to the press. He admitted that the Glacier report was not comprehensive in its treatment of the Park Service's failures—Frank said that Cole had described it as a "whitewash." Nevertheless, according to Frank, Cole said that he required the Craigheads, if asked about the Glacier investigation by reporters, to provide only information consistent with the Park Service's findings.

Judge Hauk asked Frank what Cole had said would happen if the brothers failed to do so.

"Well, he threatened that if we didn't give out what he admitted was misinformation on the grizzly report, that they would get us out of Yellowstone. But it was already evident that he was going ahead, prior to that," said the biologist.

"Say that again," responded the judge. "That if you didn't give out misinformation—"

"Yes, sir—"

"—What you thought was misinformation, but he thought was good information?"

"He admitted, in a way, that it was misinformation."

"If you didn't publicize this misinformation in the Glacier report?" continued Judge Hauk. "Which you say he admitted was misinformation?"

"Yes, sir," nodded the biologist.

"That you'd be run out of the park?" repeated the judge.

"That is correct," said Craighead.

Frank went on to tell the court that he had refused to toe the line. He told Cole that as scientists, he and his brother would tell the truth about what their research showed, not falsehoods cooked up for political reasons.

TO THE CRAIGHEADS, the temporal relationship between the superintendent's announcement that Hartzog wanted the Green Lab torn down to beautify the park, the release of the Glacier report, Glen Cole's insistence that the Craigheads support its conclusions, Frank's refusal to do so, and the lab's subsequent demolition, were indicative of the sort of process that human resources professionals call progressive discipline: a series of staged actions designed to move someone out the door. Even if the Park Service director took an interest in the demolition of a single building in Wyoming, Yellowstone had more than nine hundred other buildings and dozens of trailers used for temporary housing for park workers—yet not a single place could be found for the Craighead study. The demolition of the Green Lab was the kind of ugliness that made even the government's lawyer cringe when he allowed Frank Craighead's testimony to wander into it during cross-examination.

"So," said Spivak, "when events came to pass where there was a break in the diplomatic relations between you and the park—?"

"Our laboratory at Canyon, which I indicated was a building belonging to Yellowstone Park Company, was bulldozed and burned," said Frank Craighead.

Spivak objected. "I think that question would call for a yes or no answer," he said to the judge.

"Let him answer! Overruled," Hauk responded. "If you don't like the answers you get but you ask the questions that leave it wide-open and he drives the bulldozer right through, that's your problem, not mine." He turned to Craighead: "Go ahead and answer in your own way."

"This laboratory was destroyed, where we were working," said Frank.

"By whom?" the judge asked.

"It was destroyed by the order of the Park Service," said Craighead.

BUT THIS WAS just the beginning. As the tension between Cole and Anderson and the Craigheads grew after the demolition of the Green Lab, a revered name in national park science seemed to have sided with Cole and Anderson against the Craigheads.

"Who is Adolph Murie?" William Spivak asked Starker Leopold during Leopold's testimony in Los Angeles.

"He is one of the outstanding naturalists, who served his whole career in the National Park Service," answered Leopold.

"Is he recognized as an authority on the subject of bears?" asked Spivak.

"Yes, indeed. He did a great deal of work in Mount McKinley National Park, where he knew about bears and wolves and many of the wildlife," Leopold replied.

Four months after the Green Lab was torn down, Jack Anderson told John Craighead he was concerned about the visibility, to passing tourists, of brightly colored radio collars the brothers had installed on seven elk in the Madison River Valley. Anderson said that Adolph Murie had stopped by to see him and Cole at Park headquarters, and he had complained about all of the Craigheads' "gadgetry" on Yellowstone animals. Murie had a record of objecting to visible marking of animals in the national parks. "National parks have great value as research centers," he wrote in a 1962 memo opposing the use of the Craigheads' methods in a proposed grizzly study at Mt. McKinley National Park. "But some forms of research clash with the basic national park objectives—when they do clash, the aesthetics and spiritual values take precedence. Research technique in national parks should accord with . . . the spirit of wilderness, even though efficiency and convenience may at times be sacrificed."

The Madison River elk were a special case, as far as Anderson and Cole were concerned. Wintering in bits of snow-free meadow around geothermal sites in the Madison, Firehole, and Gibbon river valleys, they never migrated out of the park. Therefore, unlike the other bands, they weren't subject to hunting, making them as pristine a population as could be found in Yellowstone. The Craigheads had started studying them in the course of their elk research for John McLaughlin, and they secured a grant from the Atomic Energy Commission to use these elk—which were easier to find in the winter than other herds—to refine radio tracking of ungulates.

Cole was apparently not the only alumnus of Grand Teton National

Park to have been influenced by Murie. It appears that his boss had been, too. In a follow-up letter to John Craighead that almost sounded like Murie ventriloquism, Anderson wrote that conspicuous marking of animals was inconsistent with the natural-appearing park he planned to exhibit for the Centennial. And, in a passage that can only have seemed very threatening to the Craigheads' radio collar work on grizzlies, Anderson suggested that refinement of wildlife telemetry techniques might be more suited to national forests or wildlife refuges than to the purer environment of national parks.

THE SEDUCTIVE FLUSH, for the Craigheads, of fame and influence around the time of their first National Geographic television special, and the political miscalculations that accompanied it, coincided with a similar process in the life of Starker Leopold. However, in Starker's case, fame and influence afflicted him differently, inducing him to take on more work than he could possibly handle.

After 1963, as Starker rode the wave of approval for the Leopold Report, he and his colleagues on Udall's committee moved assertively on to other issues, taking on the problems with the Bureau of Sport Fisheries and Wildlife's refuge system and the government's war against predators. In 1964 Park Service director Hartzog appointed Starker to another advisory committee for that agency. In 1967, with Starker still serving both committees, Hartzog named him for the post of Park Service chief scientist, restructuring the agency's organizational chart to have Starker report directly to him. By special arrangement, Starker continued teaching and supervising graduate students at Berkeley instead of coming to Washington.

In the spring of 1968, in addition to these jobs and teaching at Berkeley, Starker was director of a Berkeley field laboratory in the Sierra Nevada, chairman of the board of the California Academy of Sciences, a consultant to the California State Board of Water Quality, and a visiting professor of wildlife management at the University of Wisconsin.

At the apogee of his power, Starker found his responsibilities unmanageable. He lost touch with his children. His son, who was in his final years of high school, graduated, joined the Marine Corps, and didn't come back. Citing his workload, Starker tendered his resignation as chief scientist effective in June of 1968 and also resigned from the Interior Department panel, but he stayed on as chair of the Natural Sciences Advisory Committee.

In the latter capacity, before the end of that month, he wrote Hartzog about the trouble with the Craigheads at Yellowstone, suggesting that a group be formed to look over the brothers' work and see how the Park Service might make use of what they had learned about bears before relations with them deteriorated to the point that the data became unavailable.

AS STARKER REACHED a crisis over his workload and the Craigheads collided with management at Yellowstone, Starker's initiative to restore fire to the national parks claimed a critical beachhead at Sequoia and Kings Canyon National Parks. In the spring of 1968 Bruce Kilgore got approval to burn 800 acres of patchy red fir forest on Rattlesnake Creek, a tributary of the Middle Fork of the Kings River. Surrounded on three sides by bare rock, the site was remote enough that had something gone wrong, visitors would never have seen it. Crews and equipment were flown in, and Kilgore and his coworkers laid out test plots and fire lines, then set fire to the woods in thousand-foot strips and allowed each one of them to burn down before the next one was started. The fire went off perfectly.

Also that year, through the work of Kilgore, John McLaughlin, and others, 60 percent of Sequoia and Kings Canyon—it would eventually be 75—was declared a natural fire zone where lightning strikes could be allowed to burn. The following year Kilgore's team moved their burning out of the remote fir forest and into the sequoia groves of Redwood Mountain. In that year, 1969, things would continue to be encouraging at Sequoia and Kings Canyon, and very worrisome at Yellowstone.

20

BEAR MANAGEMENT COMMITTEE

IF SHOOTING GRIZZLIES prevented maulings after trash separation began at Trout Creek dump in 1968, it stopped working in 1969. At Fishing Bridge in early June, five-year-old Daphne Jax of St. Paul, Minnesota, was leaving a restroom to walk back to her family's trailer when she was seized by a seventeen-year-old male grizzly that had been hit by rocks and bottles thrown by two other children. She suffered internal injuries, a broken rib, and a punctured lung. At the hospital she told reporters she was never going back to Yellowstone. The bear was killed.

Three days later, rangers tried to tranquilize and relocate Craighead research bear number 26 at Canyon campground. The brothers had captured him at the Trout Creek dump as a baby and named him the Fifty-Pound Cub. Nine years later he weighed 480 pounds, but his name hadn't changed. He regained consciousness during handling and was apparently enough of a hazard that the rangers shot him. The following week at the same campground, rangers euthanized Craighead bear number 112, along with a fourteen-year-old sow they'd previously shot and wounded.

The summer of 1969 saw more encounters between humans and grizzlies, but while humans were sometimes wounded in the scuffle, the interactions proved fatal for bears. Back at Fishing Bridge, on the night of July 9, twenty-two-year-old David Lou of Los Angeles and Michael Rock, twenty-three, of Carnegie, Pennsylvania, were pulled from their sleeping bags and bitten on the head in separate attacks. Both survived, but Lou's scalp was torn off. Craighead research bear number 183, a five-year-old male, was seen in the area after the incidents and shot, although it was not clear whether he had anything to do with the attacks. Two days later, a yearling tranquilized by rangers at Pelican Creek charged them when it woke up and was shot. The same day, also at Pelican Creek, rangers killed another bear by accident during an attempt to immobilize and relocate it. Again at Fishing Bridge, on July 21 a four-year-old male grizzly was shot by rangers with two huge rifled slugs from a twelve-gauge shotgun but was still up and walking around. He was trapped and euthanized with a lethal dose of drugs.

Meanwhile, John Craighead received a letter from Jack Anderson dated July 17, in which Anderson stated that he had decided that conspicuous marking of the Madison elk, a particularly wild and special group of animals, would end. He offered to find the Craigheads somewhere else, outside the park, to do it.

By the end of the summer two more people had been injured by grizzlies, and with the expected consequences of dump closure already descending on the park after trash separation, Anderson and Cole were resolute about completing the process. The meeting Starker had suggested, to see what could be learned from the Craigheads' research and decide how to move forward, was set for September 1969 at Mammoth Hot Springs. It was a tense, two-day assembly, chaired by Starker under the auspices of his Natural Sciences Advisory Committee. In addition to him, two other members of the committee were present: Stanley Cain of the University of Michigan, who had served under Leopold on the Interior wildlife board, and Charles E. Olmsted of the University of Chicago. Director Hartzog flew out from Washington, DC.

The Craigheads thought—perhaps not without good reason, considering Starker's statement of intent to the director the previous year—that the Park Service meant to appropriate a decade's worth of their data and hard work and then leave them in the cold. They agreed to present a talk summarizing their findings to Leopold, his committee,

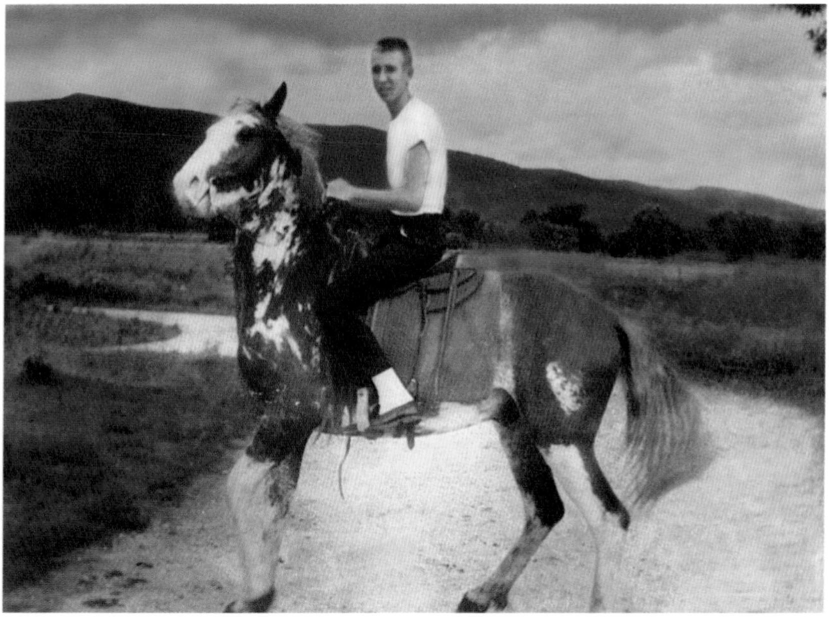

Harry Walker on Comanche, Choccolocco, Alabama, mid-1960s.
Courtesy of Carolyn Crowe

Nathaniel P. Langford, photo by William H. Jackson, 1871.
Courtesy of National Park Service, Yellowstone National Park

Stephen Zetterberg in the law library at Zetterberg & Zetterberg, Claremont, California, mid-1970s.
Courtesy of Charles Zetterberg

George M. Wright, left, and Ben Thompson in Yellowstone National Park, early 1930s, photo probably taken by Joseph Dixon.
Courtesy of Pamela Wright Lloyd

Frank Craighead, left, with goshawk, and John Craighead, with peregrine falcon, late 1930s.
Courtesy of John and Frank Craighead

Aldo Leopold, right, and Starker, camping in the Ozarks, 1938.
Courtesy of the Aldo Leopold Foundation

Aldo Leopold, left, and Olaus Murie, in 1947, the year before Aldo's death.
Courtesy of the Aldo Leopold Foundation

Starker Leopold with his father's books in his office at the Museum of Vertebrate Zoology, Berkeley, 1957.
Courtesy of the Aldo Leopold Foundation

Helicopter herding elk into trap during herd reductions in the Lamar Valley, 1960s.
Courtesy of National Park Service, Yellowstone National Park

Northern Range elk herd reduction, 1961.
Courtesy of National Park Service, Yellowstone National Park

John Craighead, left, and Frank Craighead, with immobilized grizzly in culvert trap, September 1964.
Courtesy of John and Frank Craighead

Park employees spraying diesel in first Yellowstone controlled burn, 1965.
Courtesy of National Park Service, Yellowstone National Park

Grizzly sow with Craighead radio collar and cub, 1966. The bands of colored plastic tape allowed different collared animals to be distinguished from one another.
Photograph by John Good, courtesy of National Park Service, Yellowstone National Park

Glen Cole, crouching in tan hat and raincoat, preparing grizzly for transportation in helicopter cargo sling.
Courtesy of Jim Brady

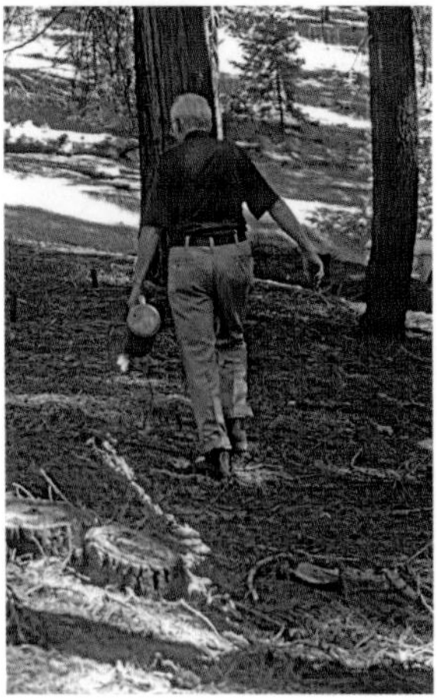

Harold Biswell demonstrating prescribed burning at Whitaker's Forest, August 1974.
Courtesy of David Graber

Harry Walker in his room at the Walker farmhouse, Choccolocco, Alabama, 1971.
Courtesy of the Walker family

Grizzlies feeding at Trout Creek dump, 1970.
Courtesy of National Park Service, Yellowstone National Park

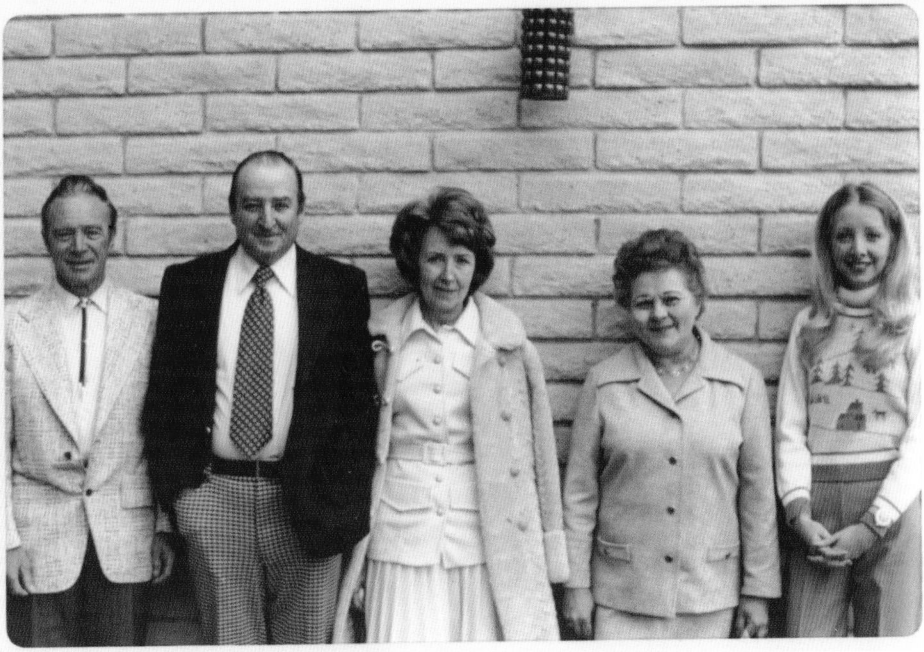

During the trial in Los Angeles, left to right: Frank Craighead, Wallace Walker, Louise Walker, Martha Shell, and Jenny Walker.

Courtesy of Jenny Whitman

Vikki Schlicht in her senior year of high school, 1971–72.

Courtesy of Vikki Schlicht

Harry Walker and Phillip Bradberry's campsite on the morning of June 25, 1972.

Courtesy of Jim Brady

Jim Brady in his office at Yosemite National Park, circa 1973–76.
Courtesy of Jim Brady

Starker Leopold, circa *Martin v. United States,* 1970s.
Courtesy of Estella Leopold

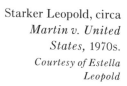

David Graber at Yosemite National Park, circa 1975.
Photograph by Howard Quigley, courtesy of David Graber

and state and independent wildlife biologists, but only if Glen Cole, Jack Anderson, and other park employees left the room. They took down all their charts and graphs before the Park Service people were allowed back in. The brothers gave a separate presentation to the entire group, including Park Service personnel, recapping their 1967 recommendations on the management of Yellowstone grizzlies. They repeated their suggestion that the dumps be shut very slowly and urged the Park Service to set out carrion in remote areas, and to close these and other locations where grizzlies gathered to visitors. And, in order to evaluate the effects of dump closure, they recommended that intensive study and radio tracking of bears continue.

AFTER THE SEPTEMBER meeting, even before Starker had turned in his committee's recommendations, Cole gave Anderson a draft program for rapid dump closure, ignoring the Craigheads' warning to take it slow. Cole's plan called for shutting Rabbit Creek by the following spring of 1970, with a temporary landfill surrounded by an electric fence nearby to replace it, and for closure of Trout Creek by 1971.

On October 6, Starker sent the Natural Sciences Advisory Committee's recommendations to Director Hartzog in Washington. He presented the options for fast and slow dump closure but didn't come down one way or another, leaving the decision to Anderson and Cole. Based on how much the bears had been stirred up by the reduction of food at Trout Creek, their exodus to Rabbit Creek and the campgrounds, Starker expected trouble when both dumps were closed. He recommended strict attention be paid to reducing the availability of food and garbage in the campgrounds, and he called, as had Adolph Murie in 1943, for study of electric fences and other barriers to protect people in campgrounds. He endorsed the Craigheads' recommendation for continued tracking and monitoring of bears to evaluate the effects of shutting the dumps. Of the bears marked and radio-collared by the Craigheads, a sizable proportion had been captured at Trout Creek dump. With the closure of that dump, those bears were the ones most likely to show up in campgrounds. For public safety alone, it made sense to keep tracking them.

But Anderson and Cole did the opposite. They directed rangers to remove radio collars and visible markings from any bear they handled, to make those bears a natural-appearing population for the Centennial.

On cross-examination, Stephen Zetterberg pressed Starker to explain the logic of this.

"Would you have approved the continuation of the radio monitoring of the bears?" he asked.

Starker responded that he and other members of the Natural Sciences Advisory Committee had no objection to continuing to monitor bears that were still carrying radios when the dump closure went into effect. But Zetterberg knew Starker's approval of continued monitoring was more than passive. He had advised Anderson and Cole to do it.

"That was one of the procedures you recommended. Is that not correct?" he asked Starker.

"Yes, so far as there were instrumented bears in the park," Starker answered.

"And marked bears?" Zetterberg pressed his point.

"Yes," replied Starker.

"And it would have been good, at least for the literature, to continue to monitor and keep track of these marked bears. Is that not correct?" asked Zetterberg, knowing full well that this wasn't what had happened at all. "Would you consider it important enough to cut that off, in order that during the Centennial, you wouldn't have any marked bears? Would that be a good biological reason for not continuing these studies?"

Starker sidestepped the question, replying that all of the bears had permanent markings. But he knew that these tattoos and small metal ear tags could not be used to distinguish individuals at a distance.

Zetterberg wasn't diverted. "My question was: Is it good procedure to start removing collars and to remove tags, in order that the bears would look spruced up for the Centennial? Is that worth the price of bad figures and bad results in terms of recordkeeping?"

"Oh, clearly you lose some data," admitted the professor, again skirting the question's broader implications. "If that is what you are implying, I certainly agree with you."

Starker had also incorporated in the bear management recommendations the Craigheads' proposal for carrion bait stations to draw bears away from people during dump closure. But once again Cole and Anderson rejected the idea. They saw it as a step backward from a natural park.

Zetterberg read aloud a passage from Starker's committee report recommending the carrion plan.

"You agree with that recommendation, do you not?" he asked Starker.

"Yes," replied Starker. "That was one of the possibilities, and we recommended it."

"And that came from the Craigheads, did it not?"

"Yes," Starker responded.

Zetterberg had put him in a difficult position. As a defense witness, Starker was there to shore up the government's case, and yet the more Zetterberg pushed, the more he exposed the chasm between what Starker had recommended—in line with the Craigheads on at least two counts—and what Cole and Anderson had actually done.

To drive home his point that the dump closure had been carried out with greater concern for a rapid retreat from influence over nature than for sound science, public safety, or common sense, Stephen Zetterberg called Frank Craighead back to the stand as a rebuttal witness.

"From the standpoint of human safety, does the carrion feeding have any importance?" Zetterberg asked Craighead.

"Yes, it does," Frank answered.

"What is the importance?" asked Zetterberg.

"The importance is that carrion strategically placed in areas away from campgrounds would attract and hold grizzlies during the period of high visitor use in Yellowstone, and this was verified by experiments that we conducted in prior years," answered Craighead.

"And that was not done, as I understand it?" the attorney responded.

"No, it was not done," said the biologist.

"I have no further questions, Your Honor," said Zetterberg.

FOUR DAYS AFTER Starker Leopold mailed his recommendations to Director Hartzog, Marian, on whom the Craigheads had installed the first large-animal radio collar in history, was killed on the north shore of Yellowstone Lake. With food reduction at Trout Creek, the brothers noticed a potentially deadly pattern. Some bears that had been regulars at the dump for years but had stayed away from areas where people were present, suddenly expanded their ranges to include campgrounds. Female number 39, for example, who had never been known to visit developed areas, suddenly doubled the size of her range and was trapped and relocated five times at Lake Village and Bridge Bay.

Marian herself had been radio-tracked for eight consecutive years and had never been known to enter a campground or other area frequented by people. On October 10, 1969, she was spotted around Lake Village with her two yearling cubs from 1968. At 7:30 a.m. on October 13, the Lake subdistrict ranger went to check a culvert trap in the maintenance yard and found two grizzlies inside it and one of Marian's yearlings outside. He darted the yearling in order to relocate it with the others, but the dose of Sucostrin wasn't quite enough to put the animal down. He was preparing another dart when Marian charged out of the forest at him, apparently in defense of her yearling. The ranger drew and fired his .44 magnum revolver, hitting Marian right between the eyes.

21

FIREHOLE

BY 1970, HARRY Walker had become a familiar face at the Smokehouse, a brick-fronted billiards hall with front windows looking out on the main business street of Anniston, half a block from the white marble courthouse where Assistant US Attorney Spivak later took depositions from the Alabama witnesses. He was a grown man now, twenty-three, six foot one, long and lean, all muscle and gristle, with a ready smile that showed his big, even, white teeth. He dressed in blue jeans, tight white T-shirts that his mother kept bleached, and plaid shirts. He liked to shoot pool, but the Smokehouse was also part of a whole mosaic of side jobs he cobbled together for extra income. Although he worked long hours on his father's farm, the dairy didn't clear enough to pay him a full salary. So in addition to the Smokehouse, at various times Harry worked the night shift at a pipe foundry, spent a couple of weeks on an oil rig off the Gulf Coast, and crewed on a shrimp boat out of Mobile. Based on his experience with farm machines, the National Guard trained him as a heavy equipment operator, and he brokered this skill into yet another job running a backhoe for an Anniston construction company. When the mail carrier on the Walkers' rural route was looking for a substitute for vacations and sick days, he asked Harry. Harry applied and got that job, too.

Harry's life was laid out before him. Like the cows that had worn their habitual trails in the red clay from the lower pasture to the milking barn, Harry would mark the hours of his life treading the same paths his daddy did, and at least for now, he would have to work those other jobs as well.

AT YELLOWSTONE THAT spring of 1970, a newly promoted thirty-year-old subdistrict ranger was sent down from Mammoth Hot Springs to take over day-to-day operations at Old Faithful Village. Jim Brady was not a big man, but he had played college football and he exuded physical confidence. His manner was thoughtful, polite, and deferential. He had a way of making people feel like they had his total attention when they talked with him, and among his subordinates he had a reputation for being preternaturally cool in emergencies.

Brady had grown up in San Bernardino, where the smog drifting east from Los Angeles met the dust and tumbleweeds of the Mojave Desert. His father was a brakeman for the Santa Fe Railroad there, and when Brady was fourteen his old man was hit by train and put out of work. As the family slipped into poverty, Brady discovered football, which earned him an athletic scholarship to Humboldt State University in the coastal rain forests of Northern California. In addition to lumber mills and muddy scrimmages, Humboldt was known for turning out foresters and wildlife biologists. Brady majored in botany and zoology and married his high school sweetheart. By the time he graduated, in 1962, he was working as a seasonal park ranger and very much wanted to be a permanent one. That summer his wife gave birth to their first child at Yellowstone National Park.

Brady's specialty within the Park Service—as opposed to a scientist such as Glen Cole or a ranger naturalist like Lloyd Parratt, whose job was to give out information to visitors and teach people about nature—was to make sure the coming and going of millions of visitors went smoothly, and to see to it that they went home in one piece. In the early 1960s, when Brady came on the scene, the Park Service was diversifying its traditional fare of awe-inspiring scenery with new parks where the emphasis was on pure outdoor recreation. Brady got his first permanent job in 1963 at one of these, Lake Mead, the reservoir formed by the Hoover Dam on the Colorado River. Just outside of Las Vegas, Lake Mead National Recreation Area was a twenty-four-hour bacchanal of alcohol, high-powered boats, firearms, and assorted other weapons. It was a gladiator school for young law-enforcement rangers. By the time he transferred back to Yellowstone in November of 1967, Brady had been shot at three times.

About the first thing they found for him to do when he got back to Yellowstone was fly around the Lamar Valley as an observer in a helicopter, chasing elk through the snow into winged fences and corrals that by then were referred to, not so affectionately, as McGee traps. After the beating the Park Service had taken in Senator McGee's hearings that March, they weren't shooting any elk, and, as Glen Cole implemented natural control, it was the last time they would give them away—984 of them that winter.

From the helicopter, beneath a ceiling of snow clouds, the Lamar Valley was a study in blue, gray, and black—and orange, where a shaft of low sun stole under the weather to light a distant peak. It had been snowing, and the icy wind had blown the powder off the ridges and into the gullies and low spots, where it settled in hundreds of parallel paths generations of elk had engraved into the mountainsides in their desperate search for sustenance. The elk trails stood out as white lines across the stony slopes, like the contours of a topographic map. The helicopter shook as the pilot swooped back and forth, herding strings of elk through the deep snow in the gullies. After the mayhem in Las Vegas, for Brady this was like coming home, and from the tiny bubble-canopied craft the immensity of the valley and the white elk paths were beautiful, and it all looked convincingly like wilderness.

BRADY MADE A good account of himself in 1968 and 1969 in the North District at Mammoth Hot Springs, and in 1970 he was made the Old Faithful subdistrict ranger. Old Faithful Village is located along the

Grand Loop Road in the valley of the Firehole River, in the western part of Yellowstone. The meadows surrounding it are bounded on the east by the timbered ridge of Observation Point and on the west by the dark lava bluffs at the eastern edge of the Madison Plateau. In those days the village contained some four hundred buildings, the greater number of them tiny brown-painted guest cabins and other small structures. The ruling presences were the two large, historic hotels on the north end, facing a semicircular plank causeway around an aperture from which, at intervals of just over an hour, the Old Faithful geyser erupted over a hundred feet in the air. Beyond the geyser, paths and boardwalks extended north into the Upper Geyser Basin and up the rounded white slope of Geyser Hill. The hill and the valley in the distance were dotted with clouds of steam rising from geysers and hot pools.

The Grand Loop Road had been constructed by Army engineers in the early days as a dirt wagon track. By 1970, it was a 142-mile strip of asphalt twenty-two feet wide connecting Old Faithful to the other clusters of hotels and cabins, stores, cafés, visitor centers, marinas, and campgrounds in the park. These features constituted less than 5 percent of Yellowstone's total area but absorbed over 95 percent of the rangers' energy. A project to divert the road around Old Faithful Village was completed for the Centennial summer of 1972, but when Jim Brady took over in 1970, on an average summer day between eleven in the morning and three in the afternoon, from six hundred to eight hundred cars, trucks, buses, and motor homes rolled right through the village. At intervals of about sixty-five minutes, when the geyser went off, drivers of passing cars would stop in the middle of the road, everyone would pile out with their cameras, and the Grand Loop would become gridlocked in both directions. There was no getting a patrol car or an ambulance through, and as one of his first administrative actions Brady put in a request for a motorcycle. Vehicle fires, forest fires, lost children and pets, fistfights over parking places or other imagined slights, spousal disputes, thefts, and the delivery of emergency messages to call home about the death of an aunt were all part of the job. Brady's life was a strange juxtaposition of quiet, sage-scented moments watching elk and bison grazing slowly across the meadows, and other moments that embodied the whole intense pageant of human existence, from birth to sudden death.

In June of 1970 he found himself stepping gingerly toward the edge of a pool of boiling hot water in the Upper Geyser Basin with a rope around his waist, held by another ranger, and carrying a bucket of strong soap. A nine-year-old boy walking through the geysers with his family had stumbled off the boardwalk and fallen into the pool. His mother and father heard the splash and turned around to see him surface with a shocked expression, then sink and die. He was boiled until the meat fell off his bones. It had been known since the turn of the century, when it was done for the entertainment of tourists, that soaping the hot springs caused them to bubble more violently. This action caused what was left of the boy to rise to the surface, where Brady and his rangers skimmed about eight pounds of the remains for burial.

With all of that, the second stage of Anderson and Cole's closure plan made things twice as bad at Old Faithful that year. The bears displaced by food reductions at Trout Creek awoke that spring to find workers bulldozing dirt over Rabbit Creek and installing an electrified fence around a temporary landfill at Nez Perce Creek, a couple of miles north. There were no power lines there, so a generator ran twenty-four hours a day, and a maintenance worker was assigned to monitor it. Alone out there at night, he witnessed huge grizzlies getting sparked off the fence and, as Brady understood it, was so unnerved by the experience that he refused to go back. Meanwhile, in a cooperative effort with the State of Montana to keep grizzlies off garbage along Yellowstone's boundaries, the town dump at West Yellowstone was closed that summer. However, as Stephen Zetterberg forced Starker Leopold to admit during his cross-examination, bears did not respond by suddenly switching to an all-natural diet.

"June of 1970 was exactly two months long," Brady remembered. Brady and his rangers spent their days dealing with people and their nights dealing with grizzlies. Scattered in the pines on the south side of the village was the housing for Park Service and concession employees. At night, grizzlies wandered the unlit gravel roads there, past the dormitories, houses, and trailers, looking for something to eat. So Brady and his rangers were setting culvert traps, tranquilizing the bears, and relocating them.

Toward the Fourth of July weekend in 1970, a trap was set right in front of Brady's trailer. One night he was awakened by the sound of the steel gate clanging shut. Luckily, his wife and three little girls were up at Mammoth Hot Springs at the time. It was about two in the morning. He got up and looked out the front window to where the porch light cast a pool of light around the trap. Sure enough, it was sprung.

Grizzlies mate between May and July, and Brady knew that around that time, if you trapped a female, there might be a male nearby. He got on the phone and woke up Mike Warren, a bear management ranger, and asked him to get in his patrol car and drive through the housing area to see if there were any other grizzlies around. Warren drove around, spotlighting the woods, for about fifteen minutes, then radioed an all-clear and pulled up in front of Brady's trailer. Brady came out to meet him, leaving the trailer door ajar as the screen door sprung shut. The two of them peered into the trap and saw a sow they'd handled before.

Suddenly Warren said, "Run!"

Both men sprinted for the safety of the trailer, dove underneath the porch railing, and tumbled right through the closed screen door, demolishing it. Brady slammed the trailer door behind them as a big, male bear attacked the porch in frustration and then went over to the trap with the female inside and started banging on it with its paws.

Brady got out a dart pistol, sat down at the kitchen table, and loaded a dart with enough Sucostrin to put down a big bear or instantly kill a human being. His hands were shaking with the adrenaline. As he finished loading the pistol he accidentally discharged it. The dart missed Warren and lodged in the ceiling. Brady loaded another, walked over to the door, cracked it open, and darted the big male with the first shot. The bear ran off into the darkness.

Brady and Warren got into Warren's car and drove around the housing area looking for the bear. Time went by, long enough for the animal to have been immobilized and then come to again. Finally they saw a light on in a small travel trailer and heard a woman screaming. The bear was up against the side of the trailer, trying to turn it over. Brady darted it again. This time it went straight down. The two rangers went back to anesthetize the trapped female bear and towed the trap containing her over to the second bear. They woke up a couple of other rangers and tried to drag the big male into the trap, but he was too heavy. So they went to get a pickup with an electric winch, snaked the cable through the trap past the sow, winched the male in next to her, and slammed the gate. The six-year-old male had been transplanted twice before and was sent to a zoo. Brady isn't sure what happened to the sow.

That was how it was at Old Faithful in the summer of 1970. Anderson and Cole had ordered dump closure to proceed, and Brady was determined to make it through the year without a serious bear incident at Old Faithful. He almost did—or maybe he actually did. We may never know, because amid the daily chaos and all the good deeds of the rangers that summer, key evidence thought to be fragmentary human remains was mishandled and lost, and although Jim Brady could not have known it, the old grizzly that would later kill Harry Walker passed through his hands.

FRANK CRAIGHEAD LEARNED about the macabre discovery along the Firehole River near Old Faithful over a month before it happened, when in the fall of 1970 he struck up a conversation at a store in West Yellowstone with a young man who turned out to be an off-duty seasonal ranger.

"Would you tell the court the conversation you had with Ranger Schroeder at that time?" Zetterberg asked Craighead on the witness stand.

"Well, the first time was the evening of October 10, 1970, when I inadvertently started talking with Gerald B. Schroeder, seasonal ranger," responded Craighead. "I then realized that he had more than average knowledge about the situation in Yellowstone, so I asked him who he was, and he told me. Through him I first learned of a possible mortality in the camp along the Firehole River." At that point in the conversation, Frank testified, he removed a notebook from the breast pocket of his shirt and began taking notes as Schroeder continued his tale.

"He said that in the Old Faithful area in 1970, there was probably or possibly another human mortality, but this apparently had not been recorded and not been released to the public."

Schroeder had told Craighead that an abandoned camp had been located, apparently torn up by a bear. Cans found there were mouthed and squashed but not punctured, said Frank. Sleeping gear had been dragged a considerable distance away from the camp, and a tuft of human hair had been found.

"There were indications that possibly a human had been killed at this camp, but no other evidence was apparently found. The camper never showed up, and the camp was removed by the rangers," Frank testified. Not long after that, he told the court, an old bear with broken teeth had been trapped at Old Faithful and flown eighteen miles into the backcountry. "Certainly a bear moved only eighteen miles away, in my experience in transplanting them and tracking them by radio, would readily come back. So there is a good possibility that the bear that disturbed this camp could also have been the bear that was in the Walker boy's camp," Craighead concluded.

IN THE COURTROOM, Assistant US Attorney Spivak was a slow, methodical sort of man given to staring abstractedly into space when he was formulating his questions. His suits were of good quality but worn and out-of-date. He transported his reams of case folders from his office on the courthouse's top floor to the courtroom in a shopping cart. On the afternoon of February 19, 1975, he led Jim Brady through a plodding recital of the geography of Old Faithful, asking the ranger to point out, on an aerial photo, locations pertinent to the events leading up to the death of Harry Walker.

On the morning of February 20, Brady took the stand again for cross-examination by Zetterberg, who wanted to talk about this possible mortality that Frank Craighead had brought up.

"Mr. Brady, state your name again for the record," said the judge.

"James M. Brady."

Brady wore his olive wool dress uniform, a gray shirt with epaulets, green tie, gold badge, tie tack, and nameplate. His wide-open, almost startled-looking, piercing blue eyes contrasted with his habitually deadpan, poker-faced expression. Zetterberg showed Brady a page from the log kept next to the two-way radio at the front desk of the Old Faithful ranger station. In a column on the left side were the times of various events in twenty-four-hour military notation. To the right of them were handwritten notes about what happened. Zetterberg directed Brady's attention to an entry made at 1651 hours, or 4:51 p.m., on August 23, 1970.

"Now, it says here, 'Sixteen fifty-one, 23 August, 1970, David Hamilton, O.F.L.' What is that?" he asked Brady, pointing to the initials.

"Old Faithful log, apparently," responded the ranger.

"Old Faithful," Zetterberg agreed, and continued. "It reports: 'The visitors at the Upper Fire Hole Bridge are gazing—' "

Here Zetterberg paused theatrically, as if trying to make out the entry, which he knew very well.

"—and it looks like, 'are gazing at a scalp on a blanket.' "

THE HUMAN SCALP is an upholstery, stretched thin over its bony substrate, and it was not unknown for it to be peeled off in a big flap during a grizzly attacks. So the report that had come in about a scalp was a plausible one, and one of Jim Brady's rangers and a seasonal firefighter were dispatched to investigate. Arriving at 5:36 p.m. near a bridge where the Grand Loop Road crossed the Firehole, the rangers radioed that indeed there was a blanket on which lay a scalp with pieces of flesh, crawling with maggots. They collected the blanket and remains and brought them back to the ranger station.

Stephen Zetterberg called the judge's attention to an entry made in the logbook by a ranger named Stu Orgill at the time the evidence reached the ranger station.

"Your Honor, may I point out that at 1736, a person that might or might not have been Stu Orgill wrote in the log: 'Re: 1651, It is a scalp with pieces of flesh, maggots. Pictures taken and pieces collected under the rear of the building.'"

Storing flesh, or any remnant of a body that was not completely dry, in a plastic bag at ambient summer temperatures outdoors under the back of a ranger station was not something a properly trained policeman or evidence technician would have done. Without freezing or refrigeration, the moist environment of a plastic bag speeds putrefaction and decomposition. But the rangers probably considered the material too objectionable to bring inside.

Today, collection and preservation of evidence is taught at the National Park Service's basic law enforcement academy. However, for rangers in 1970, there was no academy. They learned on the job, as Brady had at Lake Mead, and even as many policemen and sheriff's deputies did outside the national parks, at that time.

On August 24, 1970, the day after the scalp was found, eight or nine rangers and firefighters, supervised by Jim Brady's boss, the West District ranger, went back to the scene and made an intensive search of the area. Another clump of hair of either human or animal origin was located and brought to the ranger station.

Jim Brady remembered looking at the blanket and other material and directing a ranger to drive them to the Law Enforcement Office at park headquarters in Mammoth Hot Springs. After the evidence arrived there on August 31, eight days after the initial find, it was mailed from the Law Enforcement Office to an FBI crime lab. But when it got there, an inventory by the laboratory technician of the items in the package did not include a scalp.

"And the records of the FBI show that what was sent was the blanket," said Zetterberg, connecting the dots for the judge. "And the report from the blanket refers to hair taken from the blanket. There is no reference to the scalp. Now, where is the scalp?"

"Is that a question for me?" asked Brady from the witness stand.

"That's right," said Zetterberg, pointedly.

"Well, Mr. Zetterberg, I would think—"

"I don't want your philosophy, Mr. Brady!" said Zetterberg sharply. "I want to know if you know where that scalp is!"

"Let's let him answer," said Judge Hauk. "He is trying to answer."

"I believe it is in Mammoth," said Brady.

"On what basis do you base that belief?" Zetterberg pressed him.

Brady explained that the Law Enforcement Office was the central repository for all evidence. Anything sent to the FBI would have been sent from there, and if it was returned by the FBI after analysis, it would be retained there.

"Wait, just a moment," said Zetterberg. "You said it was sent to the FBI, but the report there does not say it was sent to the FBI. The report says, 'Blanket Q-l sent to FBI,' and there is no reference there to any maggots and there is no reference to a scalp. And you have got a week in between. And what can maggots do to a scalp in a week?"

"I see," said Brady. "Well, to the best of my knowledge, Mr. Zetterberg, the blanket and the scalp, which were kind of stuck together, were sent."

"To Mammoth?" asked Zetterberg.

"Correct," answered Brady.

"From you? And that's the last you knew of it?"

"Yes, sir, as far as I recall."

ABOUT THREE WEEKS after the discovery of the blanket with the scalp, a park visitor reported finding an abandoned camp in the same area. Ranger Tom Cherry was dispatched to investigate. Upon arrival, he and his partner found the camp's furnishings torn up and strewn around. In contrast to the account that Frank Craighead had gotten from Gerald Schroeder, Cherry maintained that cans of food at the camp had been punctured by bears' teeth as opposed to mashed without holes. The rangers cataloged seventeen items of personal property and searched the area. No sign of the camp's occupant was found. In the following weeks Cherry attempted to match the abandoned camp with missing person reports, but nothing came of it. If someone was dragged out of camp and eaten by a bear along the Firehole River in August of 1970, no other part of him or her has ever been found.

22

THE TEMPTATION OF STARKER LEOPOLD

IN THE SUMMER of 1970 when the Rabbit Creek dump closed and the human scalp—or whatever it was—found near the abandoned camp on the Firehole River went missing from evidence, what the Craigheads predicted would happen in the event dumps were rapidly closed did indeed happen. Bears wandered across roads, into campgrounds, and into neighboring national forests and settlements looking for food. A grizzly was shot illegally at West Yellowstone. Three days later one was hit and killed by a car along the Firehole River. Another was injured in a collision with an automobile and was shot on the scene. Several succumbed to overdoses of veterinary drugs during attempts to relocate them. In the Targhee National Forest, along the park's west boundary, the Forest Service issued a grazing permit to a sheepherder in the middle of grizzly habitat. Four bears were shot there to protect sheep. In that one year, fifty-seven grizzlies were lost to the Yellowstone population in human actions. Added to deaths from natural causes and what could be expected to be a lower birth rate due to the loss of females, the attrition in the grizzly population seemed to the Craigheads an arrow pointing to extinction.

Ten of the dead were Craighead research bears, with whom the brothers and their assistants had spent hundreds of hours. The researchers knew these animals in the affectionate way we know our horses, our dogs, and our cats. They knew their appearances and personalities, their life histories, their cubs, and their cubs' cubs.

Bear number 164, for example, was captured and radio-collared at Rabbit Creek in October of 1963. Like Marian, she was not initially given a name. Also like Marian, she became part of an effort to use radio collars to track grizzlies to their secret dens to learn about the behavior and physiology of the half of a grizzly's life spent in winter torpor. That autumn the researchers followed number 164 through thick timber and snowstorms—at one point, sixteen miles in twelve hours. In early November she rewarded them with their first discovery of a den. They named her "Lucky." She stayed in the study for nearly seven years. In 1970, after Rabbit Creek closed, she drifted north out of the park and was shot by a hunter.

Bears that were relocated repeatedly and came back were given away to zoos or euthanized. It wasn't as if the rangers didn't give them a chance. In August of 1970, they killed a bear that had been relocated three times in that year alone. As Frank Craighead testified in the Walker trial, the park wasn't big enough to put grizzlies beyond their capacity to navigate back. There had been discussions about flying some of them into remote wilderness areas in Idaho or Montana. In the spring of 1970, Glen Cole contacted Maurice Hornocker—who had completed a doctorate on mountain lions under John Craighead and was now a professor at the University of Idaho—about the idea. Cole asked Hornocker to assist him with moving grizzlies to the Selway-Bitterroot Wilderness. Cole wanted to do it quietly, to reduce controversy. Hornocker turned him down. He didn't want any part of something that was both sneaky and potentially dangerous.

In July of 1970, Cole, Anderson, Hornocker, the Craigheads, and representatives of state fish and game departments met at a resort just outside the west boundary of Yellowstone to discuss relocating eighteen Yellowstone grizzlies into Idaho. The effort never got off the ground, because of the states' desire not to be saddled with Yellowstone's most difficult bears. Later, a three-agency agreement between the Park Service, the State of Montana, and the Forest Service provided for translocating bears between the park and neighboring national forests, which didn't comfort the Craigheads much, because both Montana and Wyoming still had legal grizzly hunts.

AS GRIZZLIES CONTINUED to meet their fates during the spring and summer of 1970, Starker Leopold prodded Glen Cole to explain his self-regulating elk theory. Without wolves and mountain lions, Starker just didn't see how the elk population would limit itself. Yet his challenges to Cole on the subject remained friendly, even avuncular, and he doesn't seem to have taken Cole and Anderson on about the way the dump closures were going. Why was Starker not more critical?

One reason may have been that Starker wasn't aware of how crazy things were getting at Yellowstone. Jack Anderson had set two objectives for bear management: to get bears back on natural foods and to reduce the numbers of human injuries. This was a politically skillful piece of work. Old dumps with earth bulldozed over them were de facto evidence that bears must be eating something else, and dead bears didn't cause injuries. Under those criteria the effort was an immediate success, even as the campgrounds descended into chaos and dozens of bears were killed. It is a well-known pitfall in the professional lives of the powerful that their subordinates and coworkers don't want to give them bad news about projects they are involved in.

Cole admired Starker, the famous Berkeley scientist, and wanted his approval, and he kept Starker informed that things were going well. Starker didn't have time to check the reports. He had too many other irons in the fire.

After the crisis of 1968 that led him to resign as chief scientist, Starker's resolve to cut his workload didn't last. He was, even in late career, passionately interested in his field, flattered by each honor, and attracted to each chance to do something good for nature. In 1970, when he was elected to the National Academy of Sciences, in addition to teaching and supervising graduate students, he was consulting in Africa for the Tanzania National Parks, and was about to be named to an advisory group to the President's Council for Environmental Quality. He was still director of UC's research station in the Sierra Nevada and president of the California Academy of Sciences, and he sat on a committee for the conservation department in Missouri, where he'd done his doctoral research. Meanwhile, he was constantly called upon to review and comment on other scientists' papers.

Beyond sheer busyness, another factor compromised Starker's critical eye on the Park Service. There was something about being prominent and associating with powerful people that Starker really liked. He sat on the Advisory Committee on National Parks with the wife of a former president, Lady Bird Johnson. He shared a trip to Mexico with aviator General Charles A. Lindbergh. He could telephone the director of the Park Service and the superintendents of America's greatest national parks at their homes and call them by their first names.

At Yellowstone in particular, Starker's chummy relations with the staff were cemented by his annual fly-fishing trips on the park's blue-ribbon trout streams. Each year, in late summer or early fall when the snow had melted off the high peaks and the water settled down, Starker came to fish the Madison, the Firehole, and the Yellowstone with Glen Cole. Cole was a fine fisherman and agreeable company, and he had keys to all the gates. Starker often brought a friend from California, and sometimes Cole's wife, Gladys, and their girls would accompany the party, and they'd have a bottle of wine with their picnic lunch on along the riverbank. In fact, Starker was in Yellowstone fishing with Glen Cole the day the scalp was found along the Firehole River. There is no indication that he was ever told about it.

Perhaps it was for these reasons that, despite disagreeing with Cole's science, Starker came to support him over the Craigheads. When the dump closures started in earnest in 1970, Anderson and Cole did not implement the Craigheads' notion of setting out carrion bait in the backcountry and excluding the public from these areas, as Starker suggested in his October 1969 committee recommendation. But they did make a public show of accepting Starker's advice to continue intensive research on grizzlies during closure, inviting John Craighead to submit a proposal to do it. But it was a Potemkin village, and Starker was complicit in it. After the Craigheads' proposal was sent to Starker for comment. Starker wrote to Jack Anderson on June 22: "I find nothing specifically objectionable about this research proposal, except for the fact that it originates with the Craigheads who have been so singularly uncooperative with other projects in the past." He went on: "I don't see any particular reason to extend an unpleasant and difficult relationship . . ." Anderson wrote back on June 26 that he did not plan to approve the Craigheads' research, principally because of the reasons Starker had pointed out.

Barred by the superintendent from marking or collaring bears and elk, John Craighead wrote to Jack Anderson, reminding him that one of the purposes of national parks was to support scientific research. Anderson wrote back that this was true, but only if the studies didn't reduce aesthetic values, as would putting visible markings on over two hundred bears and more than a thousand elk. By now John could see he was getting the bum's rush. What he didn't see was that Starker was behind it. John quarreled with Cole until finally, in August of 1970, Cole sent him a terse letter: "Frankly, John, I give up," he wrote, "and I would request that you deal directly with Superintendent Anderson."

BY THE TIME the FBI lab's report on the Firehole River blanket came out that October, Jim Brady had laid off his seasonal rangers for the winter, and he and Tom Cherry were the only law enforcement rangers left at Old Faithful. It was the time of hyperphagia, when bears eat voraciously to build fat for hibernation, and ten to twelve hungry grizzlies were hanging around Old Faithful Village at the same time. One sow with cubs stuck her giant head through the open door of a trailer in which a woman was cooking dinner. When Brady arrived, the bear and her cubs were sitting nearby. Brady tranquilized them and flew them to the backcountry. Another old grizzly had been repeatedly seen late at night around a fenced area at the back of a café, where food wastes and garbage were stored. Jim Brady set a culvert trap and captured the bear in mid-October.

The café grizzly was a twenty-year-old female with broken teeth—

the same one Frank Craighead would later remember in connection with the camp at the Firehole River. She did not have either a Craighead ear tag or a number tattooed in her armpit, which the Craigheads did in case the bear's tag came off. Some of the techniques the brothers had introduced—immobilization drugs, dart rifles, and ear tags—had been adopted by the rangers, but in keeping with the no-visible-markings policy, the Park Service used small metal tags that could not be seen or read from a distance. But the old bear didn't have one of these, either, and because she was unmarked, Brady deduced she had not been in trouble with people.

Park policy allowed rangers to try moving bears before killing them, and Brady decided to give her a chance. He drugged her with a jab stick—a hypodermic with the flat end of the plunger glued to the end of a dowel—through one of the ventilation holes in the trap. When she settled down he took her out of the trap and punched one of the little Park Service tags through her ear, bearing the number 1792. Then he put her back in the trap and towed it to where a helicopter could land, put her in a cargo net, and sling-loaded her eighteen air miles to a place known as Gibbon Meadow, where he unloaded her and watched her from a distance until she woke up and walked away. And as far as anyone knows, grizzly 1792 stayed out of trouble until a year that was an anagram of her tag number: 1972.

23

NATURAL REGULATION

IN THE SPRING of 1971, the Walkers' pasture was in its usual exemplary condition, and you could see it in the glossy black-and-white coats of the cows dotting the luxuriant grass from the bottomland up the hill past the farmhouse and barns, and down either side of the driveway to the Old Downing Mill Road. And you could taste it in the milk. One loyal customer swore he had been close to death, and drinking Wallace's milk had restored him to health.

A Choccolocco farmer once remarked that Wallace Walker had picked the right land. He had good grass, and if you have good grass, you have good cows, and if you have good cows, you have good milk. More than that, Wallace was a superb husbandman, and having picked a good pasture, he knew how to regulate it.

By the end of the sixties the Walkers had 59 milk cows and about 350 acres of their own and rented land in pasture and feed. In the spring, Wallace and Harry plowed and planted hay and corn to keep the cows through the next winter. In the summer, they'd cut and bale the hay and pile the bales in an orderly, fragrant stack in the hay barn. When the corn ripened, they'd run the harvester through it, which piled the corn in a wagon. Then they'd grain-shovel the cobs from the wagon onto a mechanical conveyer pulled up against the corn barn, and the cobs would tumble into a golden pile inside. When haying or harvesting required more hands, Harry recruited some of his friends, and his mother would feed them big lunches of hamburger steaks, biscuits, and gravy.

Harry's younger sister would steal admiring glances at the boys with their shirts off. No one who is still alive has forgotten those times: the lowing of the cows and the mesmerizing buzz of cicadas in the steamy heat of the afternoon, and the scent of new-mown hay mixing with the hot, oily smell of the farm machinery and the sweat of the men when they came in to wash up at the kitchen sink for supper.

That life, however, was rough on men's bodies. Wallace had been plagued by back problems. He'd undergone surgery when Harry was twelve, which wasn't all that successful, and he was seldom free of pain. Over time his knees grew arthritic and he became terribly allergic to the corn he grew for cattle feed. He'd come in from working in the cornfield with a nosebleed and masses of red welts on his forearms. In 1970, the corn blight began in Florida and swept through Alabama, and on top of the allergy, that put an end to the Walkers' corn-growing days. Wallace and Harry continued producing hay, but from then on Wallace had to purchase feed. Unpasteurized milk of the kind he had distributed in the late forties and the fifties was now illegal in Alabama, and Wallace sold most of his product to a wholesale creamery, although, being known for the healthfulness of his product, he kept a small retail business going for customers who brought their own jugs. But in general, Wallace spent the rest of his career squeezed between the price of feed and fuel, the work's cost to his body, and the wholesale price of milk.

Then there was Harry's elbow. It had been broken when he was kicked by a horse at the age of fourteen. Around 1970, someone in the family made a home movie of the summer haying. In it, Harry walks away from a tractor hooked to a wagonload of baled hay, massaging his right elbow with his left hand. By July of 1971 the elbow was so painful he went to Anniston General for X-rays. A doctor injected it with cortisone and recommended that Harry take up sedentary work. In November he was back at the emergency room, now complaining of pain in his upper back and neck, possibly from favoring the elbow during hard, physical labor. He was given a cervical collar.

BACK IN SEPTEMBER of 1957, when Jim Brady started college at Humboldt State, on the Redwood Coast of Northern California, an eighteen-year-old kid from the Sacramento Valley, Doug Houston, did, too. Not long after Houston's arrival, he was walking down the hall of his freshman dormitory when he smelled a steamy odor of fish. He followed it to the open door of a dorm room occupied by eighteen-year-old Jim Brady. Brady, who was dirt poor and there on a football scholarship, was making soup from a salmon or steelhead he'd hooked in a local stream, in a coffee can supported by two bricks and heated by an inverted steam iron. They became fast friends.

When Brady went to work as a ranger, Houston went to graduate school. He did his doctoral work on moose at the University of Wyoming and was hired as a wildlife biologist at Grand Teton National Park by Glen Cole. In 1970, Cole brought him up to Yellowstone, where he joined two other scientists working under Cole: William Barmore, the park's range management specialist, and one of Starker Leopold's former graduate students, the bison biologist, Mary Meagher. The four of them were the deepest science team Yellowstone had ever had, a realization of Starker's desire to see groups of PhDs deployed in national parks. When Houston arrived, Cole was tied up with the struggle over bears, and he asked Houston to work on elk. Houston was bright and an extremely hard worker. Published in 1982 as *The Northern Yellowstone Elk,* his research remains one of the most exhaustive studies ever conducted on that animal.

Houston helped Cole reformulate his 1967–68 policy of natural control into a hypothesis known as "natural regulation."* Natural regulation proposed that elk herds are self-regulating units, which even in the absence of predation could not grow beyond the limits of their habitat. It was a remarkable hypothesis, considering some sixty years of concern over the threadbare condition of the Northern Range.

---

* During the controversy over natural regulation, its defenders argued that it was not a policy but working hypothesis. However, in its original iteration natural control was expressed as a philosophy of how elk should be regulated and how people should see them, not as an experiment.

NATURAL REGULATION AND *natural control* were not terms Houston or Cole invented. They were present in the work of David Lack, the British ornithologist who during World War II figured out that the ghost signals on British radar were flocks of birds. Lack's radar work was only a prelude to his more well-known contributions to science. His biographer called him the father of evolutionary ecology, the study of how species evolve in relation to one another and to their physical environment. Before and after World War II, Lack worked in the classic Darwinian setting of the Galápagos Islands. An illustration from his 1947 book, *Darwin's Finches,* showing how species of Galápagos finches developed radically different beaks to exploit their different food sources, would be reproduced in almost every basic biology text published for the next sixty years.

Even more influential at the time Cole was coming into his career was Lack's 1954 work, *The Natural Regulation of Animal Numbers.* Lack believed that wildlife populations were—within a natural range of variability—inherently stable and didn't expand to the extent that they theoretically might because of what he called "natural controls." These controls were "density dependent," meaning that as the number of animals in a given area went up, they competed with one another for food and places to live and passed on diseases and parasites more easily, resulting in lower survival rates of young and greater mortality generally. Lack dismissed predators as a limiting factor. In fact, he said, predators were limited by their food supply, the available prey.

One of the problems of studying nature is that other than a fragmentary historical and paleontological record, we have only the present world, used hard by our forebears, on which to base our deductions of how nature works; and, by extension, how it is supposed to work. After the murderous rampage against predators of the late nineteenth and early twentieth century, American biologists entering the field in the 1950s and '60s really did see a world lacking large carnivores, in which ecosystems were regulated from the bottom by the amount of plant food available to herbivores, and not from above, by predation. Since before these biologists of the '50s and '60s were born, deer and elk had been going through cycles of wild population growth followed by mass die-offs. It was, in a very real sense, the new normal.

Perhaps wildlife irruptions really *were* normal, thought Yellowstone biologists Cole, Houston, and Meagher, who were reading the New Zealand wildlife biologist Graeme Caughley. In 1970, Caughley published a widely cited paper questioning Aldo Leopold's account of the dire consequences of the explosion of deer on the Kaibab Plateau. If any illustration was reproduced in biology and environmental studies texts nearly as often as David Lack's, of Galápagos finches, it was a graph Aldo Leopold published in 1943 showing the rise and precipitous fall of deer on the Kaibab Plateau. Leopold's Kaibab deer population curve looked like a tidal wave: it went up to a sharp peak, and then it went straight down.

Caughley differed. His paper included a more pleasant graph, reflecting his observations of the introduction of a wild Asian goat, the Himalayan tahr, to New Zealand. In Caughley's analysis, when an animal exploited new habitat, its population grew rapidly, then started to decline. But, in contrast with Leopold's doomsday curve, Caughley's curve leveled off happily in the middle, with a series of smaller ripples as the animal came into balance with its food supply. Cole, Houston, and Meagher thought Caughley's model of a new animal like the tahr coming into balance with its environment might also apply to a native animal, like elk, that had been depleted in the nineteenth century and was now repopulating Yellowstone. Having gone through their crashes, maybe elk were poised to enter Caughley's happy, late stage of equilibration, if only they could be left alone.

But even more interesting to Cole, Houston, and Meagher was Caughley's challenge to the very idea of overgrazing. Who said that chewed-down vegetation was a sign that something was wrong? This misapprehension harkened back to the fact that range management, as practiced from the 1928 Northern Range study at Yellowstone forward, had been developed in agricultural husbandry and adapted to wildlife management, bringing with it certain assumptions that Caughley said applied to responsible agriculture, but not to nature. The point of agricultural pasture management was to place on grazing land something less than the maximum number of animals it could potentially support, in order to preserve its capacity to produce an economically reliable yield of meat or milk, year after year. Caughley distinguished this number of grazing animals, which he called the land's "economic carrying capacity," from its "ecological carrying capacity," which could be expected to be variable and look a little messy sometimes, but this did not mean something was wrong.

In 1971, Houston and Cole established what they called a "testable hypothesis" for natural regulation of the northern elk. They predicted that the Northern Range population, estimated to be 7,500 at the time, would stabilize without any help at 6,000 to 9,000.

THE IDEAS COMING out of the third floor of park headquarters at Mammoth Hot Springs came immediately under question by Raymond Dasmann, a former graduate student of Starker Leopold's who was now president of the Wildlife Society, the foremost organization of wildlife biologists in the United States. Dasmann wrote the Park Service to say that he and others in the society were troubled by what they heard about Yellowstone's management of elk and bears. Leopold, too, raised questions about natural regulation in a letter to Jack Anderson. In the spring of 1971, Cole presented a paper on national regulation at the North American Wildlife and Natural Resources Conference in Portland, Oregon, where Leopold challenged its logic in front of others.

"You really shook us up," Cole wrote to Leopold when he got home. Leopold replied, reassuring Cole about their relationship—"I don't see anything to get 'shook up' about," he wrote, telling Cole he was looking forward to fishing together at Yellowstone that summer—and at the same time repeating his objections to natural regulation. What about the exclosures, he asked. Since the time of George Wright, biologists had fenced off small plots, known as *exclosures,* to compare vegetation with and without grazing. At Yellowstone some of the fenced-in areas were so dense with aspen and willow that one could not walk through them, while around them were sparse sagebrush, puny grass, and a few forbs. "Some of the questions I asked personally in Portland are still not answered to my satisfaction," Leopold wrote to Cole, "particularly concerning photographic evidence from exclosures showing that both willow and aspens survive inside the exclosures and disappear outside. . . . It seems to me that we have to accept the fact that elk in the past and possibly even now have had the definite effect of changing the composition of vegetation on the Northern Range."

BY EARLY 1971, operating out of a trailer in West Yellowstone, John and Frank Craighead were spending less time on research than they were on fighting the government through a stream of letters and phone calls to legislators, animal welfare organizations, and academic allies, and in interviews with the media.

The Craigheads' permit to conduct bear research in Yellowstone was set to expire that year and needed to be renewed. Citing academic freedom, John Craighead refused to sign a new memorandum of understanding that required them to submit all publications and communications for approval to Cole and Anderson. On the first of February, Craighead wrote Anderson requesting an extension of the research permit pending an agreement on the memorandum of understanding. Anderson's refusal of February 9 read: "Your opposition, as well as that of Frank, to the park's program has been very forcefully communicated in numerous press releases and talks to various groups. . . . Since your opposition to the park's program is so clear and we wish to avoid continuous controversy while we are attempting to do a very difficult job, I request that you consider the studies of bears . . . completed."

THAT SPRING OF 1971, the Park Service ceased delivering refuse to the Trout Creek dump and filled in the pit. Ivan the Terrible, a grizzly the Craigheads had taken pains not to shoot there in 1966 (he'd regained consciousness as they were working on him, chased them to their car, jumped up on the hood, and attacked the vehicle as they tried to drive away) was killed by rangers at Yellowstone Lake. In August, Loverboy, so named when he snuck in to breed with Marian while a dominant male was chasing another rival away, was euthanized by rangers after reportedly chasing hikers along the South Arm of Yellowstone Lake. He had been studied by the Craigheads for nine years. In September, grizzly number 177, whom the Craigheads had known since they tagged him as a yearling in 1964, wandered into Gallatin National Forest and was shot by a hunter. By the end of 1971, forty-nine Yellowstone-area grizzlies were gone, seventeen of them Craighead research bears.

John Craighead continued to make his case in the media. In a major *New York Times* story about their battle with the Park Service that August, the Craigheads were quoted warning that the grizzlies of Yellowstone were in danger of being eliminated by the very policy that was supposed to protect them. In the Department of the Interior, a level up from the director of the Park Service, an ambitious young assistant secretary for fish, wildlife, and parks, Nathaniel Reed, took a personal interest in stopping this criticism from within his own chain of command. John Craighead's Cooperative Wildlife Research Unit at the University of Montana was a research organization within the Bureau of Sport Fisheries and Wildlife, an agency of the Department of Interior, of which Reed was assistant secretary. In November of 1971, Reed decreed that "all communications with the public, the press, the Congress, and institutions of higher learning" on the grizzly situation be cleared through his office. Craighead, who had a family to support, was muzzled if he wanted to keep his job.

At home, John brooded. He was having a terrible time sleeping. His doctor put him on tranquilizers, to be followed later by a series of other drugs for treatment of anxiety and depression. Meanwhile, in a short handwritten note to Starker Leopold in Berkeley, Glen Cole enthused: "The bear program has been going better than I even expected."

PART III

TAKE IT EASY

24

LAST STRAWS

HARRY WALKER'S MOTHER, Louise, was totally dedicated to her husband and children. However, her way of showing it could be overbearing and, at times, volatile. Sometimes she'd throw things. One of the kids' friends remembered watching Wallace dodge a plate flying across the kitchen. As a teenager, Jenny once said something disrespectful to her mother and made the mistake of turning her back. A moment later, a flowerpot shattered against the wall next to her. "She was only five foot two, had this way of cocking her eyebrow at you. She could strike terror in anyone," Jenny remembered.

Louise didn't like some of Harry's friends, who by 1971 were growing their hair and affecting a fashionably disheveled appearance. Still living at home at twenty-four, Harry felt crowded by expectations and obligations. When his resistance began, it was not limited to his relationship with his mother. In May of 1970, in Ohio, National Guard troops who had been called out to suppress demonstrations against the expansion of the Vietnam War into Cambodia fired on unarmed students, killing four. By 1971, among some of Harry's friends, a guardsman was not the most popular thing to be anymore. Then, in the summer of 1971, Harry ran afoul of his new commanding officer at the National Guard.

It wasn't uncommon for the young, part-time soldiers of the National Guard to turn up for assemblies with their hair and sideburns at a fashionable civilian length. When this happened, they were immediately sent to a barber. The milk inspector who recruited Harry in 1967 had been succeeded by a new captain with no connection to Harry's family. When Harry arrived for training camp in Mississippi in 1971, the captain sent him to get a haircut. Instead, Harry and another young guardsman visited the camp's inspector general to complain. When the captain found out, he meant to teach Harry a lesson. There was a marksmanship competition coming up, and Harry would be representing his unit on the rifle range—something he enjoyed and was genuinely proud of. The night before the match, his commander assigned Harry duties that kept him up all night. The next day he was in no condition to shoot well. The incident frustrated him. Around this time, someone in Harry's family snapped a picture of him back home in his room. He was wearing his military fatigues with his unit's shoulder insignia, but he had started growing a beard, like his friends. He also began spending the night away, at the home of John Dormon, his best friend from high school.

A MILE AND a half from the Walker farm, the Dormon place was a single-story white craftsman-style farmhouse set well back from the road up a dirt driveway. Its porch roof cast a deep shadow over its front windows, like a baseball cap pulled low over a pair of bloodshot eyes. By 1971, John Dormon's father had long since died, his mother had moved to town, and John and his two brothers were presiding over the first hippie house anyone remembers in the Choccolocco Valley. Carloads of young people came and went, some local, others with out-of-state plates.

In the spring of 1972, a classmate of Harry's and John Dormon's from Oxford High School, Phillip Bradberry, was staying at the Dormons'. He was four months younger than Harry. He'd grown up in a tiny three-bedroom cottage on a weedy hillside on the other side of the Southern Railway tracks from the better houses in downtown Oxford. When they moved into the house, Phillip's mother and father couldn't agree on which bedroom to occupy, so Phillip's mother slept in one, his father in another, his sisters in the third, and Phillip slept on the sofa in the living room. Phillip's father worked in quality control at the Oxford foundry of U.S. Pipe. He was a secret drinker. Phillip himself started drinking in his freshman year of high school. He dropped out as a sophomore. At eighteen he was drafted and sent to Vietnam, where his war was characterized not by danger or heroism but by crushing boredom. He served long shifts on perimeter sentry duty, caught a bad case of malaria, and spent months in a military hospital before being shipped back to the States and discharged. He was so emaciated that when he got off the bus in 1970, one of his sisters, who had come to pick him up, didn't recognize him and walked right past him.

After returning from Vietnam, Phillip drank, drifted, and sometimes worked at his father's plant. In 1971, he hitchhiked to California, but by early 1972 he was back and staying at the Dormons', seeing Harry when he visited. Phillip made a striking figure on the streets of Anniston or Oxford. He was over six feet tall, skinny, and wore his kinky red hair long, so that it stuck out in every direction. His friends had nicknamed him Crow, for his tendency to flap his long arms when guarding another player at pickup basketball games.

That spring, Phillip Bradberry was exactly what Harry Walker was not. Harry's life was characterized by a net of obligations and unrelenting work. Phillip was a sweet guy, but he was an alcoholic and about as unmotivated a human being as had ever been made. Harry's bedroom window at his parents' house faced the road out of Choccolocco, but he hadn't used it much. Beyond his annual training in Mississippi with the National Guard, a few visits to Mobile, and a vacation with one of his older sisters, her husband, and their children in Florida, Harry had never been anywhere. Phillip, on the other hand, was all about the open road.

SINCE 1970, HARRY had been pulling shifts, sometimes overtime, as an equipment operator in Anniston. The farm still required his work, and because he was employed elsewhere on weekdays, he caught up on farmwork on the weekends. During summer he sometimes worked eighteen to twenty-two hours a day. Wallace fretted because he couldn't pay Harry what he was worth. He felt he was living off his son's free labor. Wallace and Harry had a talk about what to do. They agreed Wallace would try to get a loan to scale up the operation so Harry could quit his other jobs and draw a salary. Wallace went to see a friend at Production Credit in town and filed an application for a $6,000 line of credit. His initial plan was to add twelve cows to the existing herd of fifty-nine.

By now Harry's elbow hurt every time he pulled the levers on a backhoe or on one of the National Guard's tank wreckers. He couldn't pick up a glass of iced tea without pain. He went to see an Anniston orthopedic surgeon, who diagnosed his condition as tennis elbow, a repetitive-motion inflammation. Again, the doctor injected the elbow with cortisone, but it didn't do any good, and on April 30, 1972, Harry was admitted to Anniston General for surgery. He was discharged with his arm in a cast on the third of May.

Harry never reported the injury to the National Guard. He just stopped showing up. A close friend later testified in a deposition that Harry had soured on military service, as well as on his new commander. After attending a weekend muster in February of 1972, Harry failed to appear for the March and April assemblies. In May, when the cast on his arm would have served as a perfect visual explanation, he didn't go either. In mid-June, after he had already left home, the guard declared him AWOL—absent without leave. At that time, once guardsmen who went AWOL were tracked down, they were turned over to the regular Army to serve the remainder of their enlistment. As Harry may have seen it, under the influence of the disaffected veteran Phillip Bradberry, that might have meant going to Vietnam.

AROUND THIS TIME Harry began talking to Wallace about taking a vacation. He just needed to get away—to get some perspective on things. Wallace approved of the idea, but Harry asked him not to tell his mother about it.

"She might try to out-talk me," he told Wallace.

On the first weekend in June, when Harry's arm was still two or three months from being healed, he got a phone call from John Burdette, the post office mail carrier. Burdette wanted to take a week off, go fishing, and work on his house. The post office allowed each carrier to designate only one alternate, so it was Harry's responsibility to help. Rural mail carriers provided their own vehicles. Some purchased old right-hand-drive post office Jeeps at auction, but Burdette's was a left-hand-drive SUV without power steering, too wide for the driver to fill mailboxes from the right passenger window. So Harry had to cross the pavement, face traffic on the wrong side of the road, and distribute the mail from the driver's side while steering with his weak arm.

On Tuesday, the sixth of June, Burdette received a phone call from Louise Walker. She was sobbing. Something had happened to Harry in the mail truck, and he hadn't come home last night. She had driven over to the Dormons' to ask after Harry that morning and was told he had left town.

When he got off the phone, John Burdette thought he'd better go and deliver the mail. He found his International Harvester Scout parked behind the Anniston post office with collision damage. Inside was a note from Harry. Someone had run him off the road, he wrote, and he promised to pay for the damage. He had borrowed $3.50 from the change fund Burdette kept to sell stamps to his customers, and he promised to pay that back, too. "I'm sorry to do you like this," the note said. "You've been good to me."

The vehicle was drivable, so Burdette gathered the mail and went out to distribute it. When he got to the Walker farm, he drove up the driveway and knocked on the door. Louise came out, very upset and weeping. It really got to her that Harry had left town without saying goodbye, but something else she'd heard that morning distressed her even more.

When Harry was nineteen, he'd married his sixteen-year-old high school girlfriend. She was terribly unhappy living with her parents, and Harry's sisters believe he married her to help her get out of their house.

The marriage lasted only six months, but Harry had kept in touch with her. In the hours before his departure he had gone to see her. Harry's young ex-wife told Louise that Harry had come to say goodbye to her forever. He said he had a deadly disease—maybe cancer—and intended to go away and die where he wouldn't be a burden to his family. When Louise heard this she was beside herself. She tracked down Harry's orthopedic surgeon to inquire about his medical records. There was no sign of any such illness in Harry's chart, the surgeon told her.

No one knows for sure what Harry was thinking when he told his ex-wife he was dying. But he was AWOL, and for the rest of his life Phillip maintained that when they left town they had talked about crossing the border into Mexico or Canada. The way Phillip put it, they were "running from the draft."

IN ADDITION TO Phillip and sometimes Harry, among the shifting cast of characters staying at the Dormon place in early June of 1972 were two young men and a young woman whose exact identities no one remembers. One was from Detroit and called himself Pee Wee. One of them had a gray 1957 Buick. The girl was from Anniston, and Harry and Phillip knew her sister. In early June they were leaving town.

Phillip later told federal investigators that it was his idea to take off with those kids in the Buick. By this time Phillip had a habit of drifting in and out of Anniston and Oxford. He would just get up one morning and walk out the door with his backpack without telling anyone where he was going. After Harry's death, when Phillip didn't respond to William Spivak's attempts to subpoena him for a deposition, the US Attorney's Office asked the FBI to track Phillip down. An FBI agent knocked on the door of Phillip's parents' house in early 1973, and Phillip's mother let him in. She said she had no idea where Phillip was. When the agent asked when he was expected back, she replied, "I'll know when he's coming home when he walks through that door."

On the sixth of June, in 1972, in the moist, green heat and cicada songs of early summer, Harry Walker left the Choccolocco Valley with Phillip Bradberry and three other kids in a gray 1957 Buick with Florida plates. They rolled north through Gadsden, past Nashville, and crossed the Kentucky line, stopping in Louisville, where they stayed overnight at the home of Harry's older sister Carolyn, whose husband had moved his family there for a job as a draftsman.

In the morning, they all went outside to say goodbye. Harry wasn't much of a hugger, but on the porch he embraced Carolyn and held her tight. When he let go of her to turn and walk down the porch steps, she noticed his eyes were wet. The other kids were standing around the Buick, and the driver got in and turned the ignition key. The engine wouldn't start. Harry looked up at his sister on the porch, grinned, and shrugged apologetically in a way Carolyn loved. The driver turned the key again and the big V8 engine caught. Harry got in and the car pulled away. That was the last time anyone in Harry's family saw him alive.

25

TAKE IT EASY

FROM KENTUCKY, HARRY, Phillip, and the three other kids in the Buick drove northeast to Kent, Ohio, where they stayed with an acquaintance of the driver. From there they followed Interstate 90 up the shore of Lake Erie to Buffalo, to stay with a friend of another of the kids in the car. They doubled back to Cleveland, then over to Chicago, where they got lost on the freeways. From there they drove west toward the Rocky Mountains.

Crossing the Great Plains, they were stopped by the highway patrol in Nebraska. Phillip had lost his wallet back home and didn't have any ID. He was taken to the police station, then released. They continued west past the wheat fields and cornfields and feedlots and the brick business districts of the little towns. The hot wind fluttered their hair through the open windows. They saw the Rocky Mountains rise up to the west, at first just a blue, wavy line above the glaring yellow plains. They got a meal in Denver, then drove toward Aspen, pulling off to sleep on a dirt road in the mountains. They stayed in a campground outside Aspen, where they met a girl who was on her way to work at Yellowstone. She said it was so beautiful, they had to see it. Harry called home from a pay telephone and his father picked up the call. He was as happy as he'd ever been in his life. He told Wallace how amazing the mountains of Colorado were, and how he hoped someday his father would see them, and he asked his father to mail some of his camping gear general delivery to Cheyenne, Wyoming.

There were hitchhikers everywhere that summer, carrying backpacks and holding up pieces of cardboard with their destinations scrawled in felt marker. Other kids picked them up, shared their food, took them home, and gave them a place to stay. The Eagles' "Take It Easy" had just come out. It was a joyful, bluegrass-infused anthem about a boy hitchhiker getting picked up by a pretty girl in a Ford. Leaving Aspen, the kids in the Buick picked up a girl hitchhiker, and she let them stay in an apartment she shared in Boulder. From there Harry and Phillip struck out on their own, thumbing north. At Cheyenne, they went to the post office and picked up Wallace's packages containing Harry's green, two-person Army Signal Corps surplus tent, pots and pans, a sheath knife, and a .22 caliber rifle.

On June 19 Harry called home again. Wallace told him that the loan to bring him on full-time at the farm had come through. Harry wouldn't have to work those side jobs anymore. "That's great, Daddy," Harry replied.

Harry and Phillip hitchhiked north through Wyoming, catching a long ride north on I-25, then I-90. They crossed the Montana line and turned west, and on June 21, the longest day in the year, they spent the night at a KOA campground where the Yellowstone River emerges from its valley into the prairie hills at Livingston. The next day, Harry hitched into town for groceries.

THURSDAY, JUNE 22, was Vikki Schlicht's day off from her summer job as a hotel maid at the Old Faithful Inn, so she drove into Livingston. The young temporary workers who staffed Yellowstone's hotels, restaurants, and gift shops quickly got sick of eating the same cafeteria food, and on their days off they'd drive into town for a decent hamburger, do some shopping, maybe take in a movie, or just leaf through the shiny magazines on the rack at a grocery store.

Vikki was eighteen, thin, and pretty, with long, wavy brown hair. She had grown up on a wheat farm in high plains along the Montana–North

Dakota line, where she'd spent the previous summer working for her father, driving a big green John Deere tractor. In her final year of high school her parents had bought her her first car, a teal-blue Ford Galaxie sedan. In the spring of 1972, Vikki responded to a newspaper ad offering summer jobs at Yellowstone. It would be her first time away from home.

Harry Walker had gotten his groceries and was standing with his thumb out by the Dairy Queen on the southern edge of Livingston, hitchhiking back to the KOA. Vikki slowed down. The boy by the road was tall, thin, and tanned, with tousled, wavy brown hair, a scruffy beard, and a nice smile. She stopped, and Harry put his groceries in the car and slid in next to her. He told Vikki he was traveling with his friend Phillip, who was back at the KOA. Vikki drove Harry to the campground, where they picked up Phillip and the two men's backpacks.

At around two o'clock in the afternoon the three of them left the KOA, driving southbound up the Yellowstone River Valley. The blue-green forests and charcoal cliffs of the Absaroka Mountains rose on their left, and to their right, the riffles of the Yellowstone River sparkled in the sun. It was the time of year when tufts of cottonwood down drift across the road from the groves of trees along the river, and the horses swish their tails in the new grass. And it was the rainy season of early summer in the northern Rockies. Ahead of Vikki, Harry, and Phillip, clouds gathered over the Yellowstone Plateau.

Phillip fell asleep in the backseat. Up front, Harry and Vikki were getting to know each other. Vikki wasn't sure she trusted Phillip, but she immediately liked Harry. He was gentle, funny, caring, and sweet, with a lovely Alabama accent. She noticed he had long, pretty eyelashes and strong arms, ropy from farmwork. The two of them intuitively understood each other. They had both been driving John Deere tractors for their fathers—a life that fewer and fewer Americans knew anything about.

The Yellowstone Valley narrowed. They passed through the rocky defile of Yankee Jim Canyon, and on the far side they saw the wall of mountains at the northern boundary of Yellowstone loom up in front of them, and they entered the little tourist village of Gardiner. On the south edge of town, where the sagebrush hills rise into the park, a ranger in the little log-cabin guardhouse at the entrance saw the employee sticker on Vikki's car window and waved her through. And because the other two occupants of the car were presumed to also be employees, not hitchhikers, they didn't get the short lecture and handful of literature warning them about scalding geothermal pools and bears that every tourist entering Yellowstone was given.

Five miles up the road, Vikki stopped at a filling station at Mammoth Hot Springs. Majestic elk with huge antlers grazed on the lawns of Fort Yellowstone, with its rows of turn-of-the-century officer's quarters, barracks, and across the street, a big hotel and shops. When the gas tank was filled, the three young people continued south on the Grand Loop Road toward Norris Junction, then through Madison Junction, and on into the canyon of the Firehole River.

Along the way they came to a place where the road was blocked by haphazardly parked cars, campers, motor homes, and people on foot, carrying cameras. Vikki slowed and snaked her way through the pedestrians. Phillip was awake now, and he saw a small black bear lumbering up the line of cars as the occupants held morsels of food out their open windows and people on foot tossed them on the ground next to the bear. He had never seen anything like it. Where Phillip and Harry came from, wildlife didn't stand around on lawns or beg along the roadside. Phillip had never even seen a bear. In Alabama, he later explained to an investigator, "What's the animals' is the animals', and what's people's is people's."

Just before five o'clock, Vikki Schlicht and her two passengers drove into the parking lot of the female concession employees' dormitory at Old Faithful Village. They all got out, Vikki opened the trunk, and Harry and Phillip shouldered their backpacks, thanked her, and headed northwest on foot, looking for a place to camp.

Had Old Faithful's campgrounds been open—they'd been closed in 1968 because of bear problems—Harry and Phillip might not have used them. Phillip later said that he and Harry didn't want to be crowded in with all the regular tourists. They'd talked about hiking into the wilderness when they got to Yellowstone. Now, they could have gone to the visitor's center to inquire about destinations and get a wilderness permit. But Harry had no desire to leave Vikki.

Although wilderness backpacking was welcomed, the Park Service prohibited camping close to roads and "developed areas" like Old Faithful, where the sheer numbers of people and the resulting sanitation problems, fire danger, and wear and tear on the countryside were prohibitive. However, outlawing it didn't stop young people from doing it. During the previous summer, in 1971, the Old Faithful rangers had issued twenty-five tickets for illegal camping, asked fifty parties to leave without citing them, and located the remains of forty illegal campfires from groups they hadn't caught. By 1972, Jim Brady periodically assigned a horse patrol ranger to ride a five-mile circuit around Old Faithful to root out illegal campers. Nevertheless, on some busy evenings when he had no rangers to assign to ejecting them, he could walk up the trail to Observation Point and see the illegal campfires twinkling in the forests all around him.

HARRY AND PHILLIP crossed the footbridge over the Firehole River and followed the gray, weathered, wooden causeway past the boiling, aquamarine pools; steaming, rust-colored trickles; hissing steam vents; and chuffing geysers on the slope of Geyser Hill. About half a mile northwest up the valley, the boardwalk took them along the base of the timbered ridge northwest of Observation Point. They looked around to see if anyone was watching, then stepped off the trail to the right, crossed the white soil around the hot pools, entered the timber, and began climbing the ridge.

A couple of hundred yards up the ridge they found a flattish clearing in the forest. It was an inconsequential place, the details of which—a group of boulders and, just downhill, a tangle of fallen limbs and three fallen trees in a triangle—would never have been recorded by anyone had things turned out differently. As soon as they set their packs down, Harry and Phillip were set upon by mosquitoes. They walked back down to Old Faithful to buy insect repellant and returned to set up camp. At about nine o'clock, Harry walked back down the valley to look for Vikki. He found her at the dorm, they talked for a while, and then Harry returned to camp. It rained all night.

26

OLD FAITHFUL

On the morning of June 23, Harry and Phillip awoke surrounded by the deep shadows of the dripping pine forest, under a platinum sky. Their tiny, primitive army-surplus canvas pup tent hadn't stopped the dampness from invading their cotton clothes and cheap sleeping bags. So they got up and picked their way back down through the woods to the boardwalk, then strolled down off Geyser Hill looking for a place to warm up.

At a snack bar at the Old Faithful Lodge, they struck up a conversation with a young waitress, who poured them free cups of coffee. They wrapped their cold hands around the mugs, inhaling the steam. The kids working as short-order cooks had messed up an order, soaking a corned-beef sandwich in griddle grease, and the waitress gave it to the two hungry Alabamans, who consumed it with relish. They thanked her and left the snack bar. For the next several hours they explored Old Faithful Village, watched the geyser go off, and wandered the trails. Harry was just killing time until he could see Vikki.

When Vikki finished her shift, she gave Harry and Phillip a tour of the Old Faithful Inn, where she was working. The lobby was a spacious atrium at the center of the building that soared seventy-five feet up to roof trusses made of varnished logs, surrounded on four sides by posts made of tree trunks supporting interior balconies with railings made of varnished pine poles. The wood glowed in the light from parchment-shaded sconces on the tree trunks and shafts of daylight from dormer windows, high above, in the roof. A wood fire flickered in a huge stone fireplace. In front of it, guests sprawled with books and magazines on an arrangement of comfortable furniture. There was nothing like it in Anniston, Alabama.

Harry wanted to show Vikki their camp, so the three of them walked back up the boardwalk across Geyser Hill. When they got to the campsite they made a fire and talked. Around midnight, Harry walked Vikki back to her dormitory. When they got there they weren't ready to part. They talked some more, about anything, about everything.

Harry told Vikki about his farm, and said he would like to take her there to see it, and to meet his family. He told her how much he liked working with the animals there. He wanted her to meet his horse, Comanche. He confessed that he'd been married once, but it didn't work out, and Vikki said that was okay. Harry asked her if she might like to see Mexico City. Vikki thought she might. The two of them shared their likes and their dislikes. Harry said he didn't like being around a lot of people, which was why he and Phillip had camped alone, up that ridge, and again Vikki said she understood. Eventually Harry bid Vikki goodnight and walked back to camp in the dark. Phillip had cooked a meal in the tent over a can of Sterno. Harry ate, and then the two of them went to sleep in the tent.

On Saturday the twenty-fourth, in one of a series of events celebrating the Yellowstone Centennial, the governor of Wyoming, Superintendent Anderson, various other dignitaries, and a crowd of tourists listened to Wyoming senator Clifford Hansen give a speech dedicating Old Faithful's new visitor's center, a large, modern building with huge windows looking out on the Old Faithful Geyser.

Harry and Phillip watched the geyser go off again, then loafed in rustic armchairs on the interior balconies of the inn, reading newspapers and magazines and watching people cross the lobby beneath them. There was no further talk of going into the wilderness. Phillip later told investigators he was just being patient, letting his friend work out his new relationship with this girl. Phillip phoned his mother in Alabama and asked her to wire him money in care of Vikki, who had plans to go to town. At around 3:30 that afternoon Harry went and found Vikki at work in order to ask for the exact spelling of her name for Western Union, and he gave her a hard time about her maid uniform. But he would have been happy with any pretense under which to see her. They arranged to meet at five o'clock at the women's dorm.

That evening Harry invited Vikki to come back to the campsite again, where he would cook her dinner. Vikki said she planned to go to West Yellowstone, so Harry and Phillip went back up to their campsite and prepared dinner over a campfire. They put a pot containing the leftovers at arm's reach in a crook in a tree to keep animals away from it, and left the rest of their groceries tucked under a tarp next to their tent. Then they walked back into the village. Vikki hadn't left the park after all, and the three of them went to the bar in the Old Faithful Inn, ordered drinks, and talked and laughed for hours.

Close to midnight, Vikki said she had to get back to her dorm and Harry walked her home. They might have taken the long way. By this point, they were holding hands when they walked. The path they followed passed through a grove of pines. Harry stopped, leaned toward Vikki, and asked if he could kiss her.

No boy had ever asked Vikki that way before. She gave him permission, and they kissed under the sheltering canopy of limbs. Later Vikki would say that it was as if she'd spent a whole lifetime with Harry in those two and a half days. And in those final minutes there existed around them a realm of possibility, of what might have been—a home, maybe a Montana wheat farm or an Alabama dairy, children. The two of them, riding tandem on Comanche or walking with backpacks around Mexico City or Calgary.

IT WAS A Saturday night in late June, and most of Yellowstone's campgrounds were full. The rangers' car radios crackled with the voice of the dispatcher in Mammoth Hot Springs, telling them to watch for a yellow pickup, the occupants of which had been seen feeding a bear around the Northeast Entrance. Another be-on-the-lookout was broadcast for suspects in a fraud committed at Old Faithful. Yet another bear had been hit by a car near Yellowstone Lake and was now reported to be crawling, dragging its hindquarters. A ranger was dispatched to shoot it. A visitor who had been involved in a collision with a Park Service pickup wanted to meet with a ranger. A vehicle description was broadcast in order that an emergency message could be delivered to its occupants; the party was located and the message delivered. The rain had swollen the park's streams, and half an hour before midnight came a report from the parents of two boys who had waded the Lamar River earlier in the day and were now stranded, wet and freezing, on the other side. A ranger crossed the river in an inflatable raft and built the boys a campfire to get them through the night, then made plans to extract them when the water went down.

After kissing Vikki goodnight, Harry rejoined Phillip and the two of them left the Old Faithful Inn to walk back to camp. The rain had let up, and a moon and sharp stars shone through black gashes between the clouds, illuminating the white barrens of Geyser Hill and the sulfurous steam drifting across the boardwalk.

They stopped. Harry lit a cigarette, and the tip glowed orange as he inhaled. Over the previous days, as they were hiking around, Harry kept singing snatches of the theme song from a 1950s television show about the American frontiersman Davy Crockett. Every kid in America knew this song. Harry had made up a new line about a bear, which he included as he belted out the song for Phillip.

Davy, Davy Crockett, king of the wild frontier!
Davy, Davy Crockett, king of the wild frontier!
Met a bear and got a great big hug . . .
Davy, Davy Crockett, king of the wild frontier!

They continued walking. On the far side of Geyser Hill they started looking for a pair of sticks they had set next to the boardwalk to mark the point where they had left it to go cross-country to their camp. They walked back and forth, peering at the ground, but they couldn't locate them, so they struck off across the geyser field and waded into the tangle of huckleberry, wild geranium, and fireweed among the trees at the base of the hill.

It was darker in the forest. They walked elbow to elbow so they wouldn't lose each other. Harry was still singing, and they were laughing and carrying on. The pale amber circle from Harry's flashlight bobbed across the scabby trunks of the pines, and then something was moving ahead of them. Harry saw it first and trained his flashlight on it for just a second, because a second was all there was. It was a bear, and it was five feet away from them once they saw it. Phillip jumped or was struck and fell, then rolled down the slope. He heard Harry scream. Phillip got up and ran. Behind him he heard Harry cry out, "Help me, Crow!" After he got some distance away, Phillip stopped, gulping air, and yelled back up the hill, "Harry, is there a bear up there?" or something similar. There was no answer, because the bear had Harry by the throat and Harry had taken his last breath and said the last thing he would ever say.

In Alabama, Harry's eldest sister, Betty, and her husband, J.C., were renting a little white sharecropper's cottage in a cotton field across the road from the Walker farm. All the Walkers were pretty tight, but Betty had a special relationship with her little brother. She was almost like a second mother to him. In the early morning hours she had a terrible nightmare. Harry was crawling toward her, dragging his backpack, and pleading, "Help me, Betty!"

J.C. woke to the sound of Betty weeping.

27

THE SEARCH FOR HARRY WALKER

WITH AROUND TWO thousand transient souls in hotel rooms, cabins, employee dormitories, trailers, and houses, at full capacity, Old Faithful Village had an overnight population equal to that of a small town. Yet from Phillip Bradberry's vantage point, half a mile away, it was a tiny island of twinkling lights in a vast, dark wilderness.

In the daytime, when Geyser Hill crawled with tourists, it seemed like a civilized place. After midnight, when the few pairs of lovers lingering on its close-in paths deserted them for their beds, the wilderness gathered in against the very edges of the lights from the buildings, and being alone out on the deserted causeways of Geyser Hill was unnerving. The boards clunked underfoot and clouds of steam drifted across the boardwalks, obscuring what was ahead or behind.

Before cellphones, when something bad occurred in the vicinity of Old Faithful at night, people ran for the twinkling lights of the inn. The main door was made of thick planks, with a latch and hinges of heavy wrought iron. Just inside it was the inn's registration desk. When there was trouble in the neighborhood, often the desk clerk heard about it before the rangers.

BETWEEN SEVEN AND nine o'clock on the night Harry died, there had been a refresher training course for law enforcement rangers at the Old

Faithful Visitor Center, featuring a speaker from the US Attorney's Office. When it was over, most of the rangers made their way to an outdoor employee party at Goose Lake. Less than half an hour's drive north of Old Faithful, Goose Lake was where park and hotel employees held their beer parties so they wouldn't bother the customers. It was three days after the summer solstice, and twilight lingered in the gray drizzle at the end of the rain. The rangers, their wives and girlfriends, the young nurses from the infirmary, and assorted dispatchers and ranger naturalists in jeans, flannel shirts, and rain slickers were out there drinking beer and grilling steaks around a campfire.

Ranger Ken Reardon was covering Old Faithful on night patrol. Another ranger, David Trickett, a thin, bespectacled teetotaler who later became president of a theological seminary, was at his residence in the village. Around midnight, the group at Goose Lake drove back to Old Faithful for an after-party at one of the rangers' residence. Tom Cherry went back to his quarters with the on-call nurse, who later became his wife. He radioed to dispatch that they would be available there by phone if Reardon needed backup on the usual things that occurred after midnight—an elderly hotel guest waking with chest pain or shortness of breath brought on by the oxygen-poor high-altitude air, a noise complaint in the cabins, or a car colliding with an elk out on the highway.

PHILLIP HAD BEEN drinking, and later he repeatedly told people close to him that he'd taken LSD that night. Somewhere behind him was a grizzly, and Harry's flashlight was lost. Phillip stumbled downhill through the forest and burst into the moonlight on Geyser Hill. He dodged the steaming pools, got to the boardwalk, and ran toward the lights. At 1:10 a.m., he crashed through the door into the lobby of the Old Faithful Inn and fell on the floor in front of the registration desk, weeping and begging for someone to help Harry.

Ken Reardon had stopped by the Old Faithful Inn on foot patrol and was warming his hands at the fireplace with two young men employed as kitchen help. The three of them gathered up Phillip, put him in Reardon's patrol car, and drove northwest up an abandoned section of the Grand Loop Road to where a footbridge crossed the river at the far end of the Geyser Hill trail. There they left the car, crossed the rumbling stream, swollen with rain, and hiked up the boardwalk to where Phillip and Harry had been leaving it to climb up the ridge to their camp.

In those days park rangers still kept their firearms in their cars, not on their waists, and as he walked up the trail, Reardon realized he was unarmed. He gave one of the kitchen helpers his car keys and sent him back to retrieve a .38 revolver from the glove box—not much of a comfort against a bear. Someone woke up Tom Cherry and David Trickett, and at around 1:30 a.m. they hurried out on foot to join Reardon, bringing with them Cherry's girlfriend, the nurse, carrying her little black doctor's bag.

Cherry, Trickett, and Reardon were unsuccessful in getting the incoherent Phillip Bradberry to lead them to his camp, and their calls to Harry up into the woods went unanswered. After about half an hour they returned to the Old Faithful ranger station, where a larger, better-equipped effort was mounted. Jim Brady and the West District ranger arrived to take command. Ranger Scott Connelly heard about the situation as he returned home from the after-party. Highly regarded by his patrol partners, Connelly judged that he'd had too much to drink to be wandering around the woods with a gun, so he went to the ranger station to see what he could do from there. Elaine D'Amico, a twenty-four-year-old former radio operator at the Old Faithful ranger station, was now assigned to help organize the Centennial festivities from Mammoth Hot Springs. Off duty that weekend and staying with friends at Old Faithful, she was awakened and pressed into service at the radio.

The rangers gathered flashlights, guns, and first-aid supplies, and the second wave of the search departed for Geyser Hill at around 3:30 a.m. All of the experienced rangers had seen multiple injuries by black bears, and because a grizzly hadn't killed anyone at Yellowstone in thirty years, a grizzly attack was not the first thing that came to mind for some of them. There was a sense that Harry might still be alive. A second nurse from the clinic in the Old Faithful Inn was awakened and sent out to assist with medical aid, if needed.

Phillip Bradberry had not been much use in locating the campsite, but he had told the rangers that Vikki Schlicht had visited it too. Rangers were sent to wake up the matron of the girls' dormitory, who pounded on the door of Vikki's room. Vikki was put in the back of a patrol car, driven up a road on the other side of the Firehole River, and marched over a bridge to where the search was being staged. To begin with, no one told her what was happening. Then Phillip mumbled that they had probably lost Harry, and she began to weep uncontrollably. She wasn't any more use than he was in locating the campsite, or Harry.

Brady had assigned the searchers to three-man teams, and for the next several hours they tiptoed through the dense pine forests, calling for Harry. Brady began to get a sense they might be dealing with a grizzly. He and others among the searchers reported hearing the characteristic jaw-snapping noises grizzlies make when they are agitated, and they were certain there was a bear close by. Young Tom Cherry had been a medical corpsman in Vietnam and had seen awful things. On edge like everyone else, he had left the safety off on his shotgun and accidentally discharged a rifled slug into the boardwalk, which startled everyone and set off a flurry of radio calls. Cherry was so unnerved that someone had to take him back to the ranger station.

EVERYONE WHO HADN'T been was sober by the time Harry Walker's body was located at around 5:30 in the morning. As the storm cleared, it had grown cold, and the wan, yellow-gray light of dawn reached into the shadowy tangle of the forest, making it easier to distinguish a perfectly still human form from a pile of deadfall. Harry was found lying on his back 165 feet from his tent in a triangle of fallen trees. A seasonal firefighter who saw him there remarked that his face looked peaceful, as if he were sleeping. His shirt was pulled up, his pants were pulled down, his belly was torn open, and the organs inside it were missing, along with his genitals. He still wore his raincoat.

Michael Weinblatt was a young seasonal ranger naturalist at the new Yellowstone Visitor Center. He later became a Harvard professor of medicine. He had been rolled out of bed around 3:30 in the morning to work for Jim Brady as a searcher. After the discovery of the body, Weinblatt was assigned to make a diagram of the death scene. His sketch depicts a trail of personal items—the broken flashlight, a rolled-up bandana Harry wore as a headband, a pocketknife, a comb, and coins from Harry's pockets—left as the bear dragged Harry away. Harry and Phillip's groceries were stored on the ground next to their tent, covered by a plastic tarpaulin. Some of the containers had been torn open and the contents spread around. A pot containing the remnants of the previous evening's supper was found where Harry and Phillip had left it in the crook of a tree within easy reach of the rangers, or a bear. Harry Walker's rifle lay unloaded inside the tent. A small-caliber weapon of this type would not have made any difference had he been carrying it.

Rangers swarmed over the hillside, measuring, photographing, and searching for further evidence, while covering each other with firearms against the possibility of another attack. At about 7:00 a.m., one of the kitchen helpers who assisted Ken Reardon and was still in the area was startled by the sight of a bear, above him at the edge of the woods. He yelled a warning, "Bear!" and fled toward the village, leaving the rangers on the scene glancing around them, nervously gripping their weapons.

Harry's body was bagged and carried down the hill to an ambulance from Livingston. Like Abraham Lincoln's funeral train going back to Illinois, the ambulance retraced Harry's route from Livingston to Old Faithful. Entering town, it passed the Dairy Queen where he had met Vikki, turned right onto Livingston's main street, and pulled up to the back door of a vaguely Spanish-style stucco building with a sign that read DUGAN'S MORTUARY.

That morning in the Choccolocco Valley, Harry Walker's sister Betty dressed and walked across the road to her parents' house to tell them about her dream. Harry's middle sister, Carolyn, and her family had driven overnight from Kentucky for a visit, and they drove up Wallace and Louise's driveway just as Betty got to the house. Everyone was hugging and talking, excited to see one another again. The phone rang. Wallace picked it up. It was the chief ranger of Yellowstone. At about the same moment, the local police drove up the Walkers' driveway to make the death notification. Somehow, the news traveled to the Harmony Baptist Church. The Sunday morning service was terminated and pastor and congregation drove over to the Walkers' to comfort them.

IN A LONG interview conducted that afternoon at the Old Faithful Ranger Station, Phillip Bradberry told an investigating ranger that before Harry's relationship with Vikki interceded to keep them at Old Faithful, it had been their intention to backpack into the wilderness. In the course of their discussions about this, somehow the subject came up of what to do if one of them died out there, and they each agreed not to tell anyone where the other's body was. Neither of them had any use for civilized funerals, Phillip explained. Thinking back on it, he thought Harry might have had a premonition of death. There was this "vibe" he kept getting from Harry, and that line Harry had added to the Davy Crockett song, about meeting a bear and getting a great big hug. Harry had been singing it, over and over, in the days before he died.

That afternoon—Sunday the twenty-fifth—Jim Brady and wildlife biologists Doug Houston and Mary Meagher spent an hour and a half setting three snares in the vicinity of the killing, two at the campsite and one where Harry's remains were found.

On Monday morning, when Brady and Ken Reardon went to check their traps, the snare where Harry's body had been was tripped but empty, and there was a grizzly in one of the two snares up the hill near the campsite. Houston and Meagher had selected a large pine around which to wrap the stainless-steel aircraft cable anchoring the snare. The grizzly had dug at the tree's roots with such energy that Brady thought that within a few more hours she might have toppled it. She was sitting quietly when the rangers first saw her, but as they approached she struggled against the cable, groaned, and snapped her jaws.

Brady was carrying a shotgun loaded with rifled slugs. Ken Reardon had a .300 magnum. Brady nodded. Reardon shot the bear dead. When they inspected the animal, Brady recognized her as bear 1792. The ear tag he had installed in 1970 was still there. They dragged the carcass down the hill and threw it in the back of a pickup truck for transportation to the state wildlife laboratory in Bozeman.

That morning in Livingston, a pathologist on contract with the federal government arrived at Dugan's Mortuary to begin an autopsy on Harry Walker. It was witnessed by members of the Livingston Police Department and a Park Service biologist, Edmund Bucknall. The body lay on a white enameled-steel embalming table with black chips on the sides from a mortician's belt buckle or a policeman's gun belt. It was smeared with blood. So rapid and overwhelming had been the bear's attack that there were no defensive wounds on the arms or hands. No sign was found of the fatal illness Harry had said he had when he said goodbye to his young ex-wife. The pathologist determined that the primary cause of death was suffocation, secondary to a crushing injury to the neck. There was an absence of the deep puncture wounds or tearing you'd expect from the teeth of a large carnivore. But upon exposing the interior of the neck with an incision, the pathologist found such powerful, toothless mashing that he could not make out individual anatomical structures.

On his way back to the park after witnessing the autopsy, Bucknall passed a Park Service pickup going the other way, carrying the grizzly's remains to the wildlife laboratory in Bozeman. There, a necropsy established that the animal was about twenty years old. Her canine teeth were broken-off, worn stumps. Human hairs were found in her digestive tract and on her claws.

ON TUESDAY MORNING, Phillip Bradberry woke from a fitful sleep in the nicest hotel room he would ever occupy. It was an upstairs room at the Old Faithful Inn, paneled in varnished pine and appointed with rustic furniture and cheerful, bright fabrics. Sunlight streamed through the drapes. Presently there was a knock on the door. When Phillip opened it, he found uniformed rangers standing in the hallway. He gathered his shabby possessions into his backpack and followed them outside.

In the parking lot, the air was cool and clear and smelled of pine trees and the rotten-egg odor of the geysers. There were people all around dressed for vacation: men and kids in shorts, T-shirts, and casual shoes, women in the same or in sensible travel dresses. The rangers put Phillip in the backseat of a pale-green patrol car. One of them slid in next to him. The other two got in front. They drove him to Mammoth Hot Springs and led him into a courtroom, where a federal magistrate stood waiting for him to tender a guilty plea to charges of illegal camping and failure to store food properly to keep it away from animals. Then they put him back into the patrol car and drove him south, then west through seemingly endless pine forests and fields of sagebrush with yellow and blue wildflowers. The rangers were carrying on amiably about football, but Phillip didn't feel like talking.

At long last they emerged into the sun-washed streets of the tourist town of West Yellowstone, with its faux-western facades and tired neon, and pulled up at the bus terminal. Phillip was deflated. He felt like nothing. The rangers bought him a ticket back to Alabama, and they all stood around to make sure he was on the bus when it left.

He had been in the park five days, but his whole life wouldn't be long enough to forget what happened there.

PART IV

HUMAN NATURE

28

MARTHA SHELL

AROUND THE TIME the separation and reduction of garbage started at Trout Creek dump in 1968, a veteran investigative journalist and crime writer, Jack Olsen, went to Glacier National Park to find out the real reasons why two women had died in the same night in grizzly attacks after fifty-eight years of comparative peace between bears and people.

"The answer was simple, of course," he wrote later, "too many humans infringing on bear habitat, and poor management practices by the National Park Service."

Olsen had been a managing editor for the *Chicago Sun-Times* and a bureau chief for *Time,* and by the late 1960s he was on the masthead at *Sports Illustrated.* In three feature articles on his investigation in *Sports Illustrated,* he skewered the Park Service for its complacency in dealing with the escalating incidents of the Trout Lake grizzly and routine feeding of bears at Granite Park. He later adapted the articles into a book, *Night of the Grizzlies,* published in 1969 and widely read since.

By the summer of 1972, critical coverage like Olsen's and five years of media battles with the Craigheads had taught the Park Service a lesson. Even before the autopsy on Harry was begun, Yellowstone's assistant superintendent told the press that Harry and Phillip "had done everything wrong," camping illegally and leaving food next to their tent to attract bears. Talking points distributed to Yellowstone employees in the days after Harry's death pointed to Phillip's inability even to show searchers where his campsite was, which triggered four hours of combing the woods to find Harry. The Park Service cast Harry's death as an aberration, precipitated by fatal mistakes by the victims, in an otherwise profitable effort to lower the number of visitor injuries by bears.

BY THE TIME Phillip boarded the bus in West Yellowstone, Harry Walker's remains were on their way back to Alabama by airfreight, which Wallace paid for from the money he'd borrowed to bring his son on full-time at the farm. The Vietnam war Phillip said Harry was running from was winding down, and the day after Phillip got on the bus to go home, President Nixon announced he would send no more draftees to Vietnam unless they volunteered. The local command of the National Guard notified Wallace and Louise they'd be giving Harry a military funeral, even though he was AWOL when he died.

At the reception in an Anniston funeral home, a framed photo of Harry stood in front of his flag-draped casket. He was buried in the cemetery across the road from the Harmony Baptist Church in the Choccolocco Valley. A bugler played taps and the flag was folded and given to his mother. The Bradberrys signed the guest book, but Louise Walker didn't remember seeing them. She had bitter words about Phillip, who had not come to see her since he got home or telephoned to tell her what had happened.

The fact was, Phillip wasn't talking to anybody. He was depressed. His detention, questioning, court appearance, and the stories in the media laid blame heavily on him as the survivor. Since his return from Yellowstone he had not even told his own mother about what had happened there.

A few days after the funeral, the telephone rang at the Bradberry home and Phillip's mother answered it. The woman on the line introduced herself as Martha Shell, of Kansas City, Missouri. She wanted Mrs. Bradberry to know a few things about the situation her son had been involved in at Yellowstone Park. It was part of a larger pattern of government misconduct and a big cover-up, the caller said. Did Mrs. Bradberry know the shameful way the Park Service had described Harry and her son in its press releases, as lawbreakers who'd made every possible mistake and paid for it? Maybe something could be done to correct this. Shell told Mrs. Bradberry she was working on getting a congressional investigation.

Martha Shell was a small, round, grandmotherly woman with neatly coifed gray hair, the wife of a successful manufacturer's representative in the restaurant supply industry in Kansas City, Missouri. Like Starker Leopold—who'd failed his Navy physical because of his profound hay fever while doing doctoral research in Missouri and might otherwise have spent World War II island-hopping in the Pacific with the Craigheads—Martha Shell's allergies changed the course of her life, and the Walkers' lives too.

Since the 1930s, when Martha and her husband Paul had married, when the ragweed bloomed in the Midwest they were nowhere to be found in Missouri. They bought a summer cabin in Estes Park, Colorado, outside of Rocky Mountain National Park, where Paul took up wildlife photography with a passion. He took thousands of color slides of hummingbirds at feeders, elk, deer, and other wildlife. By the 1960s they were making trips to Yellowstone, where they both particularly enjoyed watching and photographing bears. But after a few years their subjects seemed to be disappearing.

In the summer of 1966, a Cal Tech professor on a short vacation spent a single day driving Yellowstone's Grand Loop Road, where he reported seeing ten or twelve bears without even leaving his car. In 1972, he returned, and this time toured the park for four days. "In that entire period, in which I drove over virtually all the roads and hiked several miles of trails, I saw only three bears," he wrote in a complaint letter. When he inquired about where the bears had gone, the rangers told him the Park Service had been relocating them to the backcountry.

Upon taking over as superintendent of Yellowstone in 1967, Jack Anderson assigned Glen Cole to look over human injuries by bears. Since 1960 there had been 401, almost all of them involving black bears; only

28 had involved grizzlies, none of those fatal. To reduce human injuries and restore natural conditions for the 1972 Centennial, Cole and Anderson set out to deal with the hundreds of semi-tame black bears that begged from tourists and occasionally roughed them up along roads and in campgrounds. In a reversal of what had been a lax enforcement effort, under orders from Anderson and the chief ranger, in 1968 eighty-nine tourists were ticketed for bear feeding. Some bears were relocated, sometimes repeatedly, but they were as good as grizzlies at finding their way home. Since black bears were not endangered and there was no way then, nor has one been found since, to restore a food-conditioned black bear to shyness around people, a lot of them were just killed. Notwithstanding the bear Harry, Phillip, and Vikki Schlicht saw being fed along the Grand Loop Road in June of 1972, there was a drastic reduction in opportunities to watch and feed them, activities that had been immensely popular with the public. "I assure you that the disappointment of visitors is widespread, and sincere," a Florida attorney complained in a letter to his congressman after a bear-free visit to Yellowstone. He and the Cal Tech professor were not the only ones who took note of the change. So did Martha Shell and her wildlife photographer husband, Paul.

Less than five feet tall, Martha Shell was even tinier than Louise Walker, but even more formidable and far more worldly. In the late sixties she began asking questions about where the bears had gone, and by the early seventies she was carrying on an extensive correspondence with Park Service officials, including Robert Linn, Starker Leopold's successor as chief scientist in Washington. When she wasn't satisfied with the answers she got, by early 1971 she was corresponding with John Craighead, whose battle with the Park Service was widely reported in the media. Once John indoctrinated her on the plight of the grizzly, Martha Shell couldn't have cared less about black bears along roads. Saving the grizzly became her crusade.

"We are going to 'bust' this thing wide open, one way or another," she wrote in a letter to John, "and I don't know the meaning of the word failure when it comes to the situation with the bears."

Shell proved relentless. She kept an electric typewriter on a side table in her living room, and next to it an ashtray. She chain-smoked menthol cigarettes and clattered out a stream of voluminous and detailed letters to John Craighead, wildlife organizations, federal and state officials, and legislators whom she lobbied for congressional hearings. In addition, she began her own media campaign. She had long aspired to be a journalist, and she found her voice in the grizzly controversy. Drawing on John Craighead's extensive knowledge, she wrote a series of articles and guest editorials for publication in major newspapers and magazines.

Shell's vehement condemnation of John Craighead's enemies won his trust, and she became a sympathetic audience for his disdain for Cole and other Park Service people. After hearing Cole present on Yellowstone's bear program at a meeting of bear biologists in Canada, Craighead wrote to Shell: "I have never attended a scientific meeting where a speaker was more thoroughly humiliated. He had absolutely no data to support his position and the scientists present knew this."

A dramatic sense of paranoia—not unjustified by the American government's domestic spying and sabotage of dissident groups at the time—suffused Shell's communications to the Craigheads. By the summer of 1971 she was convinced that she was being investigated and that her mail was being opened. At one point she traveled ten blocks from her home, across the Kansas state line, which runs through Kansas City, and addressed the letter she mailed there to John Craighead's home instead of to his office to confuse those who she believed were interdicting her communications.

In early September of 1971, Shell met the Craigheads for the first time in person at West Yellowstone, where they based their research activities after the demolition of their laboratory at Canyon. Before the visit she had addressed John in her correspondence as "Dr. Craighead." After, she called him "John." That fall and winter Shell published op-ed pieces criticizing the Park Service in the *New York Times* and the *Kansas City Star.* She later wrote other articles and proposed still more to outdoor-oriented periodicals. She even pitched a short piece on the myth of menstruation and grizzly attacks for a feature in *Woman's Day* on superstitions that contributed to the suppression of women. In June of 1972, as Harry and Phillip were hitchhiking toward Yellowstone, Shell was meeting with Missouri congressman Richard Bolling and two aides in Kansas City to ask Bolling to lobby Montana senator Lee Metcalf, who had backed funding for the Craigheads in the Park Service budget, to investigate the Park Service's handling of grizzlies.

After Harry's death, Shell wrote to Phillip Bradberry's parents, sending them a copy of her *New York Times* editorial and explaining that there was more to Harry's death and Phillip's prosecution in the Yellowstone magistrate's court than met the eye. In a letter to John Craighead she explained that her overture to the Bradberrys ". . . merely seeks to enlist the Bradberry boy, Phil, in my campaign to obtain congressional hearings on the management of grizzlies in the national parks." For Shell, the victims in the attack and their families had become an instrument for saving the grizzly.

Shell called the Bradberrys repeatedly and left messages for Phillip. They went unanswered; Phillip's mother explained to Shell that Phillip wasn't talking to anyone about what happened. At the end of July one of Congressman Bolling's aides notified Shell that Senator Metcalf had no plans to convene hearings. At that point Shell called Mrs. Bradberry to get the Walkers' address and phone number, and shifted her focus from Phillip to the Walkers, under the theory that a lawsuit by Harry's family could open a second front in the war to save the Craigheads and the grizzlies.

On the first of August, Shell spoke with Louise Walker on the telephone. The next day she mailed Louise copies of newspaper stories in which the Park Service blamed Harry and Phillip for their own misfortune, along with a copy of Jack Olsen's *Night of the Grizzlies*. One can only imagine the impact on the bereaved mother of reading Olsen's graphic descriptions of Michelle Koons's and Julie Helgeson's screams as they were dragged into the woods, and the accounts of the survivors, who heard the snapping of the victims' bones in the grizzlies' jaws. When Shell spoke with Louise again a few days later, Louise told her that her whole family had read the book and had cried all day afterward. She was hurt and very angry, and she begged Shell to tell her what to do.

In a letter relating the conversation to John Craighead, Shell said she had told Louise that if she wanted to clear the situation up "with no doubt left about who was responsible," she should sue the National Park Service. Louise didn't think it was possible; she and Wallace could never afford such a thing. On the contrary, Shell told Louise "she would have absolutely no trouble in getting an attorney to handle such a suit on a percentage—and I gave her the name of an attorney in Claremont, California, who had already handled and won such a suit." The name was Stephen Zetterberg. Shell told Louise Zetterberg had gotten the family of another bear mauling victim half a million dollars.*

Louise said she'd never spoken to an attorney in her life. How would you even approach one in a matter like this? Shell said not to worry, she would coach Louise through the whole process. "Oh yes, please do," Louise replied. "I will do anything you will tell me to do, as you have been so good to us." Shell instructed her to sit tight until after a mid-September meeting of Starker Leopold's Natural Sciences Advisory Committee on Yellowstone bear management, which Shell planned to attend. While she was there she planned to have a look at bear handling records kept by the rangers. She didn't say how. Louise agreed to wait for instructions.

A family friend remembered her misgivings when the Walkers were drawn into a high-stakes argument over a threatened species in another state. "When I saw those big people come looking for the Walkers, my heart just sank. The Walkers were simple, country folks, and they had no idea what they were getting involved in."

THE INVESTIGATION OF Harry Walker's death was one of Jim Brady's final acts at Yellowstone. That summer the Park Service promoted him to the position of district ranger in charge of Yosemite Valley. The man who tapped Brady for the job was Yellowstone chief naturalist John

---

* This figure was inflated. The settlement for the Parratt family was reportedly $100,000, not $500,000.

Good, now Yosemite's new assistant superintendent. Brady went to orientation at Yosemite in July of 1972 and then returned to his final task at Yellowstone, the arrangements for the culminating event of the Yellowstone Centennial: a world conference on national parks featuring a visit by First Lady Pat Nixon.

On August 16, in Brady's Yellowstone district, Kenneth Bell, a twenty-four-year-old backpacker, and his sister, Carol Blackwood, were sitting on a log reading a map at Fawn Pass when they were attacked by a female grizzly with cubs. Bell fought back with a hunting knife and a hatchet. He believed he injured the bear, because he hit it once in the eye, and when the fight was over his weapons were bloodied and the bear retreated, making moaning noises. He and his sister were extracted by helicopter and taken to a hospital. During Bell's debriefing by rangers, the following exchange occurred:

RANGER JONES: One question: Do you know whether your sister was in her period?

BELL: No, she isn't. Why? What does that mean?

RANGER JONES: That makes a difference to the bear.

BELL: Oh, it makes them attack if the girl is—?

RANGER JONES: Not necessarily makes them attack, but it may be contributing.

BELL: No, I don't think so, but you'll have to ask her to make sure.

ON THE AFTERNOON of September 19, 1972, the weather had turned autumnal and it had been sleeting sideways in the northern Rockies. Jim Brady and other park officials in class-A uniforms stood in a receiving line to shake hands with Pat Nixon after the presidential helicopter touched down in a parking lot next to the Old Faithful Inn. Mrs. Nixon reviewed a line of rangers on horseback and had her picture taken strolling through the geysers. In late afternoon a motorcade carried her north to the junction of the Madison, Firehole, and Gibbon rivers, where the national park was supposed to have been invented in 1870, for a ceremony convening the Second World Conference on National Parks.

By the riverbank at Madison, a large audience of park managers and conservationists who had traveled to Yellowstone from all over the world had been served a drippy barbeque dinner and now huddled on folding metal chairs wrapped in raincoats and polyethylene trash can liners distributed by rangers. It was about 30 degrees Fahrenheit and sleeting when Mrs. Nixon and other dignitaries mounted the stage. A retainer held an umbrella over the First Lady's hairdo while she smiled gamely as the Secretary of Interior spoke for thirteen minutes. Then Mrs. Nixon dedicated a plaque, while on the far side of the river someone ignited a campfire doused with accelerant and the flames shone through the gloom, symbolizing the altruistic discussion of September 1870.

When that had been accomplished, Mrs. Nixon was whisked away to her hotel and the rest of a western campaign trip for the reelection of her husband that November of 1972. Although Richard Nixon would win the election, his presidency was already doomed by his involvement in a burglary at the Democratic Party offices in the Watergate Towers in Washington, DC. When reporters asked Mrs. Nixon to comment on the break-in during her trip to Yellowstone, she said that all she knew about it was what she read in the papers. But as the story unfolded with revelations of all the crimes that had been committed to get Nixon re-elected, a pervasive suspicion on the part of the American public that the government was lying and covering things up formed the psychic background of the Walker case. At one point, Martha Shell pitched an article to a magazine about Yellowstone's "Little Watergate."

ON THE DAY after Mrs. Nixon's visit, the bear management meeting that Martha Shell had told Louise Walker she planned to attend was convened at Mammoth Hot Springs by Under Secretary of the Interior Nathaniel Reed, who had muzzled John Craighead by ordering him not to speak to the press without permission and oversight from Reed's office. Starker Leopold, Glen Cole, Park Service chief scientist Bob Linn, and state fish and game officials were present. The Park Service had announced that the meeting would be closed to the public, but Martha Shell gained access with the assistance of Congressman Bolling, and the Craigheads slipped an article to reporters about an executive order by President Nixon forbidding closure of government scientific meetings, and to avoid further embarassment the Park Service relented.

Several talks were given, among them one by John Craighead. In 1971, when the Craigheads lost their permit at Yellowstone, they began doing pioneering work in another new field in partnership with Joel Varney, the young Silicon Valley engineer who had done much of the later work on their radio collars: computer modeling of wildlife populations. They loaded two sets of data on an IBM mainframe computer. One set consisted of grizzly birth rates, population structure, survivorship, and mortality before dump closure; the other consisted of those figures during closure. John told the audience at the meeting that the preliminary data indicated that if current conditions prevailed, the grizzly would be extinct in Yellowstone in twenty to twenty-five years.

The skirmishes over aesthetic issues like the bright-colored markings on animals could sometimes obscure real scientific differences that existed between the park and the Craigheads over actual management policy. Building on everything their radio collars had taught them about the grizzlies' movements, John and Frank maintained that management of the species could not be balkanized. They referred to the unit of habitat necessary to sustain the grizzly as the "Yellowstone Ecosystem," which they defined as a 5-million-acre cluster of public lands, including the park and surrounding national forests. Cole, on the other hand, saw the park boundary as a defensible line between preservation and exploitation and thought of Yellowstone grizzlies as national park grizzlies. Paul Schullery, an eminent Yellowstone historian, knew Cole and Anderson, worked for the park, and had friends among the Craigheads' opponents. When he approached John Craighead to write about the grizzlies in the eighties, he was rebuffed by the demonstrably bitter biologist. Schullery nevertheless credited the Craigheads with bringing the concept of the Greater Yellowstone Ecosystem to the fore, which revolutionized management not only of grizzlies, but of many other aspects of nature that don't obey lines on a map.

For Starker's part, with the dumps closed, he felt his job was finished and said that his committee's involvement with the process could end. John Craighead was disgusted with Starker's abdication from responsibility during a dangerous time for the grizzly and felt he had nothing more to lose by going at him. "I backed Starker into a corner and slugged him verbally during the question-and-answer period. I am sure he has never been quite so humiliated," Craighead wrote in a letter to family.

After the meeting, Glen Cole told reporters that the Craigheads' figures on the grizzly population were skewed by the fact that the bears they had studied were captured at dumps. Cole maintained that there was another, unstudied population of backcountry bears living entirely on natural foods, and that the Craigheads' dire predictions about potential extinction were entirely erroneous. "If you put garbage dump data in the computer, you get garbage dump data out," Cole told a reporter.

It seems that Martha Shell did procure a copy of the bear management log she'd told Louise Walker she planned to get. For the rest of her life she told relatives that she and her husband had broken into a ranger station at night to photograph the document. None of the rangers remembered a burglary, so if it happened, it is likely she used a key she'd gotten from someone. After the bear management meeting that September, she gave Louise the go-ahead and coached her on how to approach Stephen Zetterberg. In October Dennis Martin began the legal machinations that would result in moving the case to California. When Zetterberg eventually met Shell, he asked her, "How did you know about me, or should I even ask?"

THE WORLD CONFERENCE on National Parks was originally intended to take place at Old Faithful, but when it became clear that there would not be enough space, early sessions were convened there and the main conference was held at Jackson Lake Lodge in Grand Teton National Park, a blocky, modernist structure with walls of stark, unadorned concrete and huge windows with a panoramic view of the Tetons across the wetlands on the northeast shore of Jackson Lake. The autumn storm that had greeted Pat Nixon had cleared, the sun was out, and the aspens around the lodge were turning lemon yellow. On Friday, September 22, representatives of Yellowstone's progeny—the national parks of eighty-four nations—assembled in the lobby in front of those windows and that view. African park managers in flowing white robes chatted with English conservationists in bell-bottomed Savile Row suits and rangers from New Zealand, Chad, and Upper Volta. Starker, Glen Cole, and Jack Anderson were there. The Craigheads, whose cabins were within a half-hour drive, were not.

That day, Under Secretary of the Interior Nathaniel Reed gave a speech on what the United States was learning about how to manage national parks. In it he referred again to the cherished concept of the balance of nature. What had research in parks taught park managers? "From these studies, we are beginning to appreciate the fact that an ecosystem does not function effectively unless it is in a reasonable degree of equilibrium," he said. "The objective is equilibrium." How vital were predators to that equilibrium? Until recently, said Reed, predator control on public lands had made that question difficult to answer. It was hard to evaluate something you are constantly trying to destroy.

Starker Leopold had carried on a long campaign to stop government predator control programs, which, having all but wiped out the wolf south of Canada, continued to pour money into killing coyotes, cougars, and—when they transgressed onto ranches or public land livestock allotments—grizzlies. One 1963 article by Starker on the subject was titled "America's War on Wildlife." As chairman of the Secretary of the Interior's Advisory Board on Wildlife Management and from 1971, as an advisory board member of the President's Council on Environmental Quality, he had begun to prevail. In his speech Under Secretary Reed pointed to a February 1972 executive order by the Nixon administration—the result of his and Starker's partnership on the issue—prohibiting the notorious poison, Compound 1080, which, scattered from airplanes at taxpayer expense to kill coyotes, had destroyed countless other wild animals. In a private letter to Starker after the order, Reed had credited Starker's tireless pressure on the federal government for this victory.

Meanwhile, during that Centennial year, without wolves and cougars and notwithstanding the theories of David Lack and Graeme Caughley, the northern Yellowstone elk herd quietly reached ten thousand animals. Doug Houston was forced to revise the number at which natural regulation would cause the herd's growth to level off. Before long that number, too, would be exceeded. And as chief biologist, Cole continued to maintain that both the grizzly and the elk programs were working.

29

B-1

BY THE END of 1972, it had been amply demonstrated that when faced with the loss of one source of human food, bears' energetic efforts to exploit other sources would bring them into dangerous contact with people. That pattern had prevailed following the closure of the Otter Creek feeding grounds, culminating in the 1942 mauling death of Martha Hansen, and it repeated itself in the Yellowstone dump closures of 1970 and 1971. In October of 1972, Jim Brady arrived to take up his new post in Yosemite Valley in time to see the rich golds and bronzes on the leaves of the black oaks and the brilliant reds on the dogwoods, and the same pattern happening again, this time with black bears.

What was probably the most cogent analysis of the dump situation from inside the Park Service came from Neil J. Reid, a soft-spoken biologist at the regional office in Omaha, in response to a draft of the proposed plan for Yellowstone's dump closure circulated after the bear management meeting of September 1969. Reid suggested that as long as there was food for bears in the campgrounds, bears would be in the campgrounds, whether the dumps were closed or not. So he suggested—as had the Craigheads—that the Yellowstone dumps be let go for a while, or indefinitely, while a concerted effort was made to break the chain between bears and human food in the campgrounds. Along with secure garbage receptacles, Reid suggested that bear-proof food lockers could be installed either centrally or at each campsite, and that their use be made mandatory.

However, when the last of the Yosemite dumps was shut down in 1971, the park failed to secure all garbage receptacles. Food lockers, while conceived of by George Wright and Reid, had not yet been invented and deployed. So bears continued to feed on the stream of nutrition that ended up in the dumps; they just moved closer to its source in the campgrounds. Food storage regulations required campers to stow their ice chests in their cars, but even when compliance was good—it was never perfect—bears in the Sierra Nevada became very proficient at destroying automobiles to get food. As George Wright had written in 1932, nature equipped bears with powerful muscles and sharp claws to tear apart rotten logs to get larvae, dig up ground squirrel dens, and rip open the tough hide of a dead deer, elk, or bison, and these gifts worked just as well on car windows and door frames. By the end of Brady's first couple of months, Yosemite's figures for 1972 bear incidents were double the previous year's at 244 cases, mostly property damage. Three visitors had been injured, none seriously.

At that time, most of the handling of bears was done by the rangers, not by a limited staff of biologists. The use of Sucostrin and dart rifles the Craigheads had taught Yellowstone rangers had spread through the National Park Service, but calculation of the dose by visually estimating a bear's weight could be tricky, and calculation of the powder charge to deliver the dart took practice too. In one gruesome incident a ranger tried to dart a cub in front of park visitors, and the excessive powder charge blew the dart right through the little animal and out the other side, killing it. Furthermore, what Glen Cole had always believed, that there were two races of bears, good backcountry bears and bad, garbage bears, persisted among those at Yosemite and elsewhere who had not actually marked hundreds of bears and followed them around. This belief was coupled with the hope that bad bears could be exiled to the backcountry, where they might revert to their ancient ways instead of just walking twenty miles back to the campground.

However, even if bears stayed put when moved, which they didn't, there was a paucity of places to put them where they wouldn't cause trouble. A boom in wilderness recreation and backpacking had dispersed human food far beyond the roads. And although the Sierra Nevada had a lot of wilderness, it was linear, north to south; east to west it was even narrower than Yellowstone. The best bear habitat was at lower elevations, close to park boundaries, where the productivity of bears' favorite foods was highest. So, rangers trapped and tranquilized bears and released them as far away from trouble as they could, and the bears either plagued camps elsewhere or they were back before someone towed the culvert trap in which they had been captured back to Yosemite Valley. It was frustrating for the rangers to spend night after night tranquilizing the same bears, and in the end, with no alternative, having to euthanize them.

IN SPITE OF Glen Cole's reassurances that grizzlies were in no danger, an embrace among biologists of the Craigheads' idea of the Greater Yellowstone Ecosystem, and uncertainty about how many were left, led to a process by which the Park Service lost control of the grizzly issue. There had long been a simmering antipathy within state fish and game agencies toward the Park Service for "wasting" Yellowstone elk that might otherwise have been hunted in state jurisdiction. Now that antipathy found a new cause in the grizzly.

Harry Woodward, director of Colorado's wildlife agency and president of the International Association of Game, Fish and Conservation Commissioners, which had censured the Park Service for shooting elk in 1962, had gone to college in Colorado in the late 1930s when the state still had grizzlies. After graduation he worked in South Dakota, and by the time he returned to Colorado as head of the state's fish and game department in 1961, grizzlies were believed to be extinct there. The state had realized too late that it was losing them, and not enough had been done. Woodward became a leader of a working group within the Association to prevent the same thing from happening in the northern Rockies. He had aligned himself with the Craigheads and was agitating in Congress. At a meeting of the Association in West Yellowstone in January of 1972, Woodward told the press that the Park Service was likely to cause the extinction of the grizzly at Yellowstone. Pressured by the uncertainty about how many were left and the expectation that the Endangered Species Act would become law within months, in January of 1973, the Secretary of the Interior asked the National Academy of Sciences to conduct a review of the Yellowstone grizzly. The study got under way that year.

With the expectation of oversight by the Bureau of Sport Fisheries and Wildlife—now the US Fish and Wildlife Service—upon passage of the Endangered Species Act, and to retain some control over the inevitable, the Park Service set up an Interagency Grizzly Bear Study Team, composed of its own biologists and those from national forests and neighboring states, to conduct field research on grizzlies. Although the Park Service funded the team and its first leader was on the agency's payroll, this began a process of wresting control of the grizzly's future out of the Park Service's hands. In keeping with his naturalness policy, Jack Anderson forbade the team to mark or collar grizzlies in the park and, as a result, population estimates remained shaky. The Craigheads' modeling put the 1973 number at 139, and this figure is widely cited today. Although the Craigheads allowed for a percentage of unmarked bears, Glen Cole thought there were 350 bears in the Yellowstone population, but he had no solid data to prove it.

IN APRIL OF 1973, Stephen Zetterberg and William Spivak traveled to Yellowstone to take depositions from witnesses in the Walker case. During their death investigation the previous June, rangers had asked Vikki Schlicht whether she had warned Harry and Phillip about illegal camping and the presence of bears. She said she had. Thus she became a potential defense witness to substantiate that Harry and Phillip were aware of the danger of what they were doing, and Spivak subpoenaed her for a deposition at Mammoth Hot Springs. In a telephone conversation, she refused to come. The last thing she wanted was to testify against Harry's family. According to Vikki, Spivak told her she could be sent to federal prison for contempt of court. Nineteen years old and already traumatized, Vikki drove west across Montana. It was snowing hard by the time she got to Billings. She got as far as Gardiner, where she spent a freezing night sleeping in her car, and in the morning she followed a snowplow into the park.

At the deposition Vikki testified that she'd warned Harry and Phillip that there were bears around Old Faithful. "I told them on the way from Livingston that they had to get a permit to camp. They had asked where a good place to camp was so I told them to go to the ranger's station to check and get their camping permit."

In Alabama, the grieving Louise Walker was taken to the hospital in Anniston with chest pain and underwent a heart catheterization. Someone in the family took a photograph as the nurse wheeled Louise to the car upon her release. Louise put the photo in a shoebox with a note attached that she had written: "Me in the hospital with heart pain after my son Harry had his heart EATEN by a bear."

JIM BRADY HAD done some basic rock climbing, but the state of that sport on the gargantuan cliffs of Yosemite Valley far exceeded his abilities, and for that matter those of most of the other rangers. What in the 1950s had been an obscure form of athletic lunacy practiced by poverty-stricken drifters living in Yosemite campgrounds became a mass-market phenomenon that made a new class of wilderness entrepreneurs of some of the more ambitious among the penniless vagabonds. Yosemite climber Doug Tompkins started a little backpacking store called the North Face. Yvon Chouinard began hand-forging pitons in his parents' backyard and selling them out of his car. He went on to found an outdoor clothing company called Patagonia. Yosemite's hotelier opened a rock climbing school, managed by a former ranger who had been involved in the first ascent of El Capitan. The Yosemite School of Mountaineering's bright-colored T-shirts with the slogan GO CLIMB A ROCK became one of the park's most popular souvenirs among everyone but climbers, who eschewed them.

By 1973, what roadside bear sightings were to traffic jams at Yellowstone, watching climbers on cliffs was to traffic in Yosemite. Crowds of people were setting up lawn chairs along the roads to peer through telescopes at men and women creeping up the valley's walls like bright-colored ants. As the district ranger for Yosemite Valley, Brady presided over breathtaking rescues of climbers by volunteers and rangers and a surge of accident investigations, as neophytes tried out their skills, or lack of them. Nine months after his arrival, two competent climbers were on the first pitch of a classic route near Camp 4 when they saw a human body plummet past them and land on the talus, sixty feet below. They rappelled off to give aid. When they got to him, eighteen-year-old Brian Quinn had a heartbeat but was not breathing. Resuscitation attempts by a doctor and rangers proved unsuccessful, and he was declared dead on arrival at the Yosemite Clinic. He had been leading three other would-be climbers with a coil of clothesline hanging from his belt. It was only mid-July and this was already the ninth climbing fatality in Brady's domain that season. In their statement for the rangers' accident report, the two witnesses said that while the valley was filled with people sporting GO CLIMB A ROCK T-shirts, they would prefer that the concession stop selling them, or at least that the slogan be changed to DON'T FALL OFF A ROCK.

In order to be minimally competent in making decisions about rescues, recovering the fallen, and analyzing the mistakes that led to deaths or injuries, Jim Brady and other rangers were taking climbing lessons from Chris Vandiver, a young instructor at the Yosemite School of Mountaineering. Vandiver was a fresh-faced twenty-two-year-old with shoulder-length hair who was already a very competent alpinist. He was easygoing, quick with a smile, and so popular with the rangers that by the autumn of 1973 he was housesitting one of their government residences while the occupant was assigned out of the valley.

That September of 1973, the wife of another Yosemite climber told Vandiver she'd noticed the strong smell of decomposition at mile marker

B-1 on Highway 120, a twisting two-lane road blasted spectacularly into the dry south-facing cliffs at the west end of Yosemite Valley. Perhaps yet another neophyte had fallen to his death? Vandiver drove up there alone, parked at the stinky spot, tied off a climbing rope, let himself down the cliff, and crashed through the prickly canopy of live oaks that obscured the base of the drop-off from view from the road. When his feet touched down on the steep granite rubble, oak leaves, and sandy regolith, Vandiver found himself surrounded by a macabre scene worthy of the Renaissance painter Hieronymus Bosch, had the sufferers of Bosch's tortures been bears.

Scattered around him were what Vandiver estimated to be twenty black bears in various stages of decomposition, some on the ground, some hanging in trees. Their skin seethed with maggots. Vandiver saw a tiny cub that had been skinned, its head and paws chopped off. In the heat of September, the stench made Vandiver gag. Shaken, he climbed back up to the road and went to talk to Assistant Superintendent John Good. "There is more to this than you see," Vandiver recalled Good telling him. "Leave it alone, Chris." That Vandiver did not do.

VANDIVER HAD BEEN teaming up to do first ascents with a climber by the name of Galen Rowell, an aspiring photojournalist who had just gotten his big break and was about to join Tompkins and Chouinard in the pantheon of successful Yosemite entrepreneurs. Rowell was the son of UC Berkeley professor Edward Rowell and his wife, Margaret, the cellist who played duets with Albert Einstein Jr. Growing up in the middle of such intellectual riches, Galen did not latch on to them, exactly. His friend, climbing guide Doug Robinson, called him "the most hyper person I ever knew."

Galen liked speed and exposure to heights. As a teenager he began hanging around with Berkeley rock climbers, climbing in Yosemite, and building souped-up muscle cars. He made a living in his twenties operating Rowell's Automotive Service in Albany, next to Berkeley. He also liked taking pictures, and he had the presence of mind to take them on climbs when other people might have wanted to hold on to the rock with both hands. By 1972 Rowell had put up a hundred new routes in the High Sierra. One of his photos shows Vandiver standing casually on a ledge in the first free ascent of Keeler Needle, a thousand feet of air beneath his toes.

Rowell began writing magazine articles to accompany his photos. In the autumn of 1972, at the age of thirty-two, he decided to become a photojournalist and sold his auto shop. In the spring of 1973 he got a minor assignment to provide a couple of photos of Yosemite climbers in action for *National Geographic*. His photos were so good that by September he'd turned the deal into a cover story. In the process, Rowell had gotten into a dispute with one of Brady's seasonal rangers who, like the dismayed witnesses at the fatal fall in July, didn't like seeing Yosemite being commercialized. The ranger had tried to sabotage Rowell's relationship with *National Geographic,* the biggest break of Rowell's professional life. Rowell went to Brady and threatened to sue. Brady straightened things out, but coming out of that dispute Rowell was in a very bad mood about the Park Service.

Vandiver called Rowell at his home in Albany and told him what he'd found. Rowell saw it as a chance to get back at the Park Service, not to mention a great magazine story. He loaded his cameras and climbing gear into a 1966 Chevelle with a hot transmission and a balanced and blueprinted V8 and roared across California's Central Valley. Arriving in Yosemite, he and Vandiver lowered themselves down into the body dump, and Rowell shot a couple rolls of film.

Like Vandiver, Rowell's first instinct was to confront a Park Service official and see what he had to say. Jim Brady's boss, Yosemite's chief ranger Jack Morehead, was a friend of Rowell's parents and was on temporary assignment at the regional office in San Francisco. Rowell went to see him. Morehead told Rowell that of course he was aware of the body dump. When bears had been relocated repeatedly and had returned to invade campgrounds, rangers sometimes had to euthanize them with a lethal dose of animal tranquilizers. The body dump was a legitimate way to allow the carcasses to return to nature where the public wouldn't see them. Morehead told Rowell there had been only nine such killings in 1972. Rowell left the meeting unconvinced and went back to Yosemite to conduct his own investigation. What he needed now was an inside source. A ranger who was sympathetic to his cause directed him to Armand "Herbie" Sansum, a wildlife ranger in Yosemite's Resource Management office.

Herbie Sansum was an intelligent and a dutiful ranger who cared deeply for the good of Yosemite and its animals. He had been the brunt of teasing for his almost obsessive-compulsive meticulousness. Jim Brady, who himself kept a neat desk, remembered that Sansum's pencils were lined up in a straight row. Sansum had training in new animal tranquilizers, and it was his impression that while many of the dead bears at B-1 were intentional killings, others were victims of accidental overdoses by rangers using Sucostrin. Sansum hated sloppiness.

ON THE FIRST of November, Stephen Zetterberg and William Spivak drove up to Yosemite from Los Angeles to take a deposition from Jim Brady for the Walker case. A few days later, Herbie Sansum agreed to meet with Galen Rowell. As Rowell and Chris Vandiver remembered it, the meeting took place at the residence where Vandiver was housesitting. An autumn storm had moved into the Sierra. It was raining hard and the trees were blowing around. After midnight, according to Rowell, Sansum led him on foot through the downpour to the deserted administration building and let them in. They climbed the creaky old wooden stairs to the Resource Management office, where they began copying bear management files by flashlight. Late in life Sansum admitted to having helped Rowell get copies of the files—he was proud of it—but he denied ever letting Rowell into the Park Service offices at night.

A kitchen-table animal-rights organization in the San Francisco Bay Area, the Extinct Species Memorial Fund, operated by—in Rowell's words—a "Wagnerian" woman by the name of Ursula Faasii, had already gotten wind of the bear killings in Yosemite. Rowell joined forces with Faasii, and on November 20, Faasii and Rowell called a San Francisco press conference at which they displayed Rowell's photos of the body dump and confronted the Park Service with proof from Yosemite's own files that the rangers had executed twenty-two Yosemite bears in 1972 alone.

The pattern of bear killings was by no means limited to Yosemite and Yellowstone. At the press conference, Faasii shared with reporters a letter recently sent to Under Secretary Reed from an anonymous employee at Sequoia and Kings Canyon, charging rangers there with even more sadistic behavior than at Yosemite. The writer alleged that a Sequoia ranger had shot a bear with a .22-caliber bullet, knowing full well that it wouldn't kill the animal but only inflict a painful, festering wound. Rangers had killed twelve black bears in the Giant Forest area alone, two of them little cubs. A bear that broke into a cabin had been shot eight times with a shotgun. Bear management at Sequoia and Kings Canyon was cruel and unsystematic, wrote the whistleblower, with no attempt made to mark and identify problem bears before killing them.

By the end of 1973, as the National Academy of Sciences made its inquiry into Yellowstone's management of grizzlies and Galen Rowell's publicity about Yosemite bear killings made headlines, the management of the bear-human relationship in national parks was so demonstrably a failure that change was inevitable. As the Park Service had done when the elk issue blew up in 1962, they called on Starker Leopold.

30

THE DISCIPLE

After 1967, when he moved from the Museum of Vertebrate Zoology to the School of Natural Resources and Forestry, Starker Leopold's office was a two-room suite in the basement of Mulford Hall, a nondescript, mid-twentieth-century building on the north side of the Berkeley campus. From a stark basement hallway resounding with the sound of air handling and heating equipment, Starker's oak office door led into a pleasant reception chamber occupied by his secretary and containing her desk, electric typewriter, file cabinet, and mail slots for the graduate students. Beyond, through a half-glass door that was generally ajar, was Leopold's own office. It contained a wooden desk and chair, two chairs for visitors, and a wall of shelves containing the extensive library he'd inherited from his father. The walls of both rooms were hung with trophy heads of deer and bighorn sheep and a big, shiny fish. Each room had a window in the building's exterior wall, and because the basement was half-buried in the ground, the windowsills barely crested above the soil outside, affording a prospect of the undersides of the landscape plants.

In the spring of 1973, a tall, bearded twenty-five-year-old nursery worker and part-time journalist by the name of David Graber strolled through the door into Starker's office to inquire about graduate school. He had never heard of the Leopold Report and really had no idea who Starker was. But he had thought about getting a master's in wildlife biology, and he had heard that Starker had a reputation for finding his graduate students jobs.

Graber in no way fit the mold for a serious scientist. He had dropped out of biology as an undergraduate because the lab courses were too hard. His grades were nothing to write home about. But Graber was bright in a fascinatingly unshackled way. He had a way of framing sentences in the social, political, and biological contexts all at once. Starker had largely forsaken research for public policy, dedicating himself to communicating about complex biological issues to laypeople, administrators, and politicians in plain language. Graber had this ability.

Starker liked his mind.

Graber had grown up in Southern California, on the edge of the housing developments creeping out of the San Fernando Valley up the east side of the Santa Monica Mountains. For a while, when he was a boy, civilization ended at the end of his street. He went feral in the hills, where he saw coyotes and mountain lion tracks. The experience he then had, of watching his cul-de-sac become just another thoroughfare to a bunch of new houses, had a way of turning young Southern Californians into conservationists. His parents took him car camping at Yosemite, and Graber said that if you had asked his mother at that point what he was going to grow up to be, she would have told you he was going to be a ranger.

Graber graduated a year early from high school and entered UC Santa Barbara as a Regent's Scholar with a major in biology. Regent's Scholars were assigned a faculty mentor, and Graber's was the ecologist Garrett Hardin, author of the famous essay "The Tragedy of the Commons," which is still required reading in environmental studies courses. In one of their periodic conferences Graber confessed he wasn't sure he'd picked the right major. Hardin counseled him to transfer to Berkeley, where the Life Sciences Library alone was larger than the main library at Santa Barbara.

At Berkeley, Graber sampled every drug known to mankind, but hallucinogens did not blunt his intelligence, only expanded his thinking. He had this gift with language and thought he might like to be a journalist, so he changed his major to political science and began writing broadsides for the Students for a Democratic Society. But it irritated him when he saw his work in print larded by some would-be editor with all sorts of Maoist rhetoric, so he resigned. In 1968 he became friends with a political columnist for the *Los Angeles Times,* who invited him to become a student commentator on a public-affairs television show he was anchoring in Los Angeles. Graber became a regular on the show through the summer and fall of 1969. From 1969 through 1970, while finishing his undergraduate degree, he worked as a part-time journalist for *Newsweek*'s San Francisco bureau, occasionally reviewed books for the *Los Angeles Times,* and did volunteer work for Denis Hayes, national organizer of the first Earth Day.

In 1971 Graber flew to Europe and took a tramp steamer to Africa. He was looking for a career, and in the course of his travels he met some wildlife biologists in the African national parks. In October of 1972 he flew home to California, rented a flat by the beach in Venice, and picked up work reviewing psychology books for *Human Behavior* magazine and teaching environmental education in public schools. But he was thinking about graduate school now, and in the spring of 1973 he drove up to Berkeley to talk to Starker. When he told the professor about his interest in wildlife biology in Africa, he said just the right thing. Starker had served as a consultant on wildlife research there and was avid about the subject. The two of them wound up drinking beer at an off-campus dive and in spite of Graber's spotty résumé and fair-to-middling grades, within a couple weeks Starker offered him a position in a master's program in wildlife ecology.

GRABER ENTERED GRADUATE school just in time to be taught the beautiful, orderly Clementsian vision of the ecosystem as a "superorganism," in which the various parts—plants, animals, bacteria—have co-evolved, like our eyes, ears, limbs, and kidneys, to work harmoniously through a series of predictable stages toward a predestined end. And he finished his graduate work and entered professional life just in time to see the whole thing fall apart.

In July of 1973, in a paper titled simply "Succession," in the *Journal of the Arnold Arboretum,* two ornithologists, William Drury and Ian Nisbet, set out to refute the entire body of thought about succession. Drury and Nisbet were the same ornithologists who, a decade before, had used military radar to watch birds migrating 20,000 feet over Massachusetts at night. Drury, the senior of the two, was a research scientist for the Massachusetts Audubon Society and a lecturer at Harvard. His protégé and coauthor, Ian Nisbet, was a young British physicist and avid amateur birder. Nisbet had come to the United States to do physics research at MIT and watch birds. In the course of his birding he encountered Drury. Drury took the young Nisbet under his scientific wing, and in 1961 they decided to see what they might do with Lack's serendipitous World War II discovery, using modern military radar.

Drury, said Nisbet, had a total lack of respect for authority. He questioned everything. In the early 1970s, the two of them decided to question the sweet, harmonious vision of succession. In their 1973 paper, they investigated the claims of traditional ecology and found no evidence of complex interdependence between predictable assemblages of organisms with predestined outcomes. It was all random. From there, Drury and Nisbet reviewed the work of an impressive list of other scientists whose conclusions, taken collectively, had already totally torn down the principles supporting succession but without anyone taking much notice of it. They resurrected Henry Gleason, a contrarian early-twentieth-century American ecologist whose work had been forgotten in the enthusiasm for succession.

There is no "balance of nature," said Drury and Nisbet, echoing Gleason; no superorganism of species all holding hands and working together. There are only individual species with various tolerances for dryness or wetness, heat or cold, presence or absence of nutrients, sunlight or shade, forming in random combinations according to their individual needs. What seems to be a harmonious whole is just the aggregate of self-interest. There is no plan for it. There is nothing predictable or predetermined about ecology. Places like Yellowstone have no one preferred state; they might have all sorts of possible responses to chance occurrences. Drury and Nisbet had called out the beginning of the end of Clementsian succession. It took a while for the implications to filter into the offices of scientists working in national parks, but when they did, here was the problem: If a national park has no preferred state, what is our target or goal in restoring it?

THE QUESTION GALEN Rowell faced at the beginning of 1974 was a simpler one: If all animals in national parks are protected, under what authority were rangers killing Yosemite bears and throwing them over a cliff?

In the course of his investigation, Rowell telephoned Frank Craighead in Moose, Wyoming. "Bust it open!" Craighead told Rowell, according to Rowell's notes from the phone call. On Saturday, March 23, 1974, Rowell published an editorial in the *New York Times,* "Killing and Mistreating of National-Park Bears," in which he revealed that Yosemite rangers had executed more than two hundred bears between 1960 and 1972—a figure confirmed by the Park Service.

A few days later, Rowell got a phone call from Yosemite's chief of resource management, Dick Riegelhuth. According to Rowell, Riegelhuth admitted to everything Rowell had charged. But the good news, Riegelhuth said, was that the Park Service had come up with two solutions: they were writing a new bear management plan, and they would be conducting a scientific study of the bear problem at Yosemite.

"Except, there was no such study," recalled David Graber.

Riegelhuth called Starker Leopold. Graber summarized the conversation: "So Starker says, 'Okay, I'll save your ass, but on two conditions. One: You will stop using sucostrin. It's a cruel drug; the animals feel pain. Second: If you kill a bear, you will issue a press release saying why.'" Graber explained the second condition: "Starker had an adage for people in public service: 'If you're ashamed of it, don't do it. If you're not, publicize it.'"

Leopold telephoned Graber and asked him to come to his office. Graber walked in and sat down in one of the chairs.

"David, do you like bears?" asked Leopold.

"I don't have anything against 'em," answered Graber.

"Good. That will do. I have a study for you," said Leopold, and he laid out the situation at Yosemite.

"What if I don't want that particular study? What if I want to study something else?" asked Graber.

"In that case you can find another major professor," Leopold answered.

And so Graber, who had dropped out of biology as an undergraduate, was sent to reform Yosemite National Park's bear management program. Which, as it turned out, he was very capable of doing.

IN EARLY 1974, as the Walker case moved toward trial, William Spivak and Stephen Zetterberg arranged to begin taking depositions from their Alabama witnesses at the Calhoun County Courthouse, in Anniston, on a Monday morning in February. Without notifying Zetterberg, Spivak arrived a day early and showed up unannounced at the Walker place that Sunday. In the yard he encountered Harry's sisters Betty and Jenny. Spivak asked them whether their mother had been in contact with a certain Martha Shell, and whether Shell had sent their mother Jack Olsen's book *Night of the Grizzlies*. It's not clear how Spivak knew about Martha Shell or the book. Today, the FBI says it has no record of an investigation of Shell, but other federal agencies, such as the Department of the Interior's Office of the Inspector General, or the US Postal Service, may have looked into her activities. The Walker girls told Spivak he'd have to talk to their mother.

At this point, Louise came out of the house and stiffly invited Spivak to come inside. Spivak knew he was on thin ice conversing with plaintiff's witnesses without their attorney present, and he declined her invitation. Louise was pretty upset with the mere fact that he was there, but to a southern woman such as Louise, for Spivak to turn down her hospitality was the ultimate slight. Spivak asked Louise if Harry had been living at home at the time of his death. He was staying at the Dormons' place, wasn't he? The questions revealed Spivak's strategy to challenge the Walkers' claim for compensation for the loss of Harry's future work on the farm by questioning whether he would have come home had he lived. But Louise took Spivak's questions as an attack on the cohesiveness of her family, and, drawing herself up to her full five-foot-two, she hastened Spivak's departure from the property.

At the courthouse in Anniston the following morning, Zetterberg asked Spivak whether he might have visited the Walker farm without notifying him. "I know you were there," he chided Spivak with a chuckle. "You have mud on your shoes!"

In addition to damages based on calculations of the lifetime value of Harry's farm labor, Zetterberg's initial complaint sought to compensate his family for mental anguish, pain, and suffering. Spivak parried by filing a brief to the effect that by sending Louise gruesome descriptions of fatal bear attacks in *Night of the Grizzlies,* Martha Shell became an "intervening person" who contributed to the Walkers' pain and suffering. To substantiate Shell's role in maneuvering the Walkers into court, Spivak filed a motion before a federal judge in Birmingham, Alabama, to have Louise give up all letters she had received from Martha Shell. The judge so ordered, and on Zetterberg's advice, Louise complied.

Zetterberg counterattacked with a motion before Judge Hauk in March of 1974 to have Louise Walker excused from all further testimony, proposing that in light of her heart catheterization, it could be too painful or dangerous for her to undergo further scrutiny about her relationship with Martha Shell. Judge Hauk granted the exemption and struck "mental anguish" and "pain and suffering" from the plaintiff's complaint, leaving language to the effect that the Walkers had suffered the "loss of companionship, society, and comfort" of their son, which by itself could be adequate cause for damages. Thus Zetterberg ended Spivak's attempt to make the actions of Martha Shell—who maneuvered the Walkers into suing the Park Service in order to vindicate the Craigheads and save the grizzly—into a public issue. And although the trial was covered by the national media in 1975, Shell's role remained a secret known only to the parties, their lawyers, and the judges.

ON DAVID GRABER'S first day of work in Yosemite in May of 1974, Dick Riegelhuth, the chief of resource management, showed him around the valley. Beneath the cliffs where the white ribbon of Royal Arch Creek tumbled hundreds of feet through thin air into the gorge, they drove around the back of the Ahwahnee Hotel just in time to see a large bear climb out of an open dumpster in broad daylight. Graber took a picture of it, and for most of his forty-year career as a government biologist an enlargement of that photo hung on his office wall with a sign reading: OUR FOUNDER.

It had been two years since the closure of Yosemite's dumps. Trash cans had been fitted with tops with steel flaps like a mailbox, but the dumpsters were still not secure. Riegelhuth told Graber there were two things the Park Service wanted him to determine: How many bears did Yosemite have, and what distinguished an aggressive "rogue" bear from a natural one?

Graber reported the request to Starker. "Those aren't good questions for graduate work," the professor replied, then laid out a more Craighead-like project for Graber: "You're going to do a study of the entire ecology of bears at Yosemite." That spring Graber spent a couple of weeks with California Fish and Game biologists learning to capture black bears with the new veterinary drugs Herbie Sansum wanted the Park Service to adopt. Then, over the next five years, with a series of student and Park Service assistants, he set out to capture, mark, and record physical data from every bear he could find in Yosemite; a total of 298 individuals, more than the Craighead study in less than half as many years.

To do this, Graber slept in the morning, went to work in the afternoon, and stayed up all night. In order for him to carry firearms for self-protection, the Park Service put him on salary and gave him a week of law enforcement training, a uniform, and a badge. Early in the evenings, when people gathered in folding chairs around campfires, Graber walked through the campgrounds in his role as seasonal ranger, warning people to put their ice chests away and clean up their garbage. After midnight, when the folding chairs were empty and the campfires burned down to coals, Graber the ranger became Graber the researcher, stalking bears past the slumbering visitors, capturing them in culvert traps, and observing them once they were marked.

In the first year Graber determined that there were twenty-seven bears living in Yosemite Valley's seven square miles, four times the typical density of black bears existing on natural foods in similar terrain. In a population ecology study that followed the Craigheads' model, Graber learned that with their high-protein diet of human foods, female Yosemite bears were having more cubs, and the average body weight of both sexes was twice that of wild bears. One black bear that Graber captured weighed over six hundred pounds, over twice as much as the grizzly that killed Harry Walker.

On warm, typically rainless, summer nights in the Sierra Nevada, a fair number of visitors didn't bother with tents; they just rolled out their bedding on the ground. Sleeping on the ground in Yosemite in the mid-seventies was like sleeping in a backyard full of skittish and inquisitive, overgrown Rottweilers. Predatory attacks on human beings by black bears were almost unknown south of Canada; however, as had been true at Yellowstone, nonfatal injuries due to bears' fear, impatience, or aggression were common. "Black bears generally treat humans as dominant bears," Graber would report in his study, "including challenges of human dominance." Bears would walk up to people in sleeping bags and sniff at them, and if the sleepers sat up in alarm, the startled bear might claw and bite them or trample them as it fled. After the bad press about the bear graveyard, there was a moratorium on killing bears that summer, and 1974 became Yosemite's worst year in decades for bear problems. Bears were breaking into cars and buildings, tearing through the walls of occupied tents, and bluff-charging visitors to induce them to abandon their ice chests and backpacks. Property damage totaled over

$80,000 in 1974 dollars. Twenty-eight visitors were injured, nine in a single week in July.

AS STARKER DEPLOYED David Graber to work on Yosemite's black bear problem, developments in Yellowstone continued to redeem the Craigheads and extract management of the grizzly from the control of the National Park Service. In response to a legally binding petition from the nonprofit Fund for Animals, in April of 1974 the US Fish and Wildlife Service announced that it would study the grizzly bear in the Lower Forty-eight states for listing under the 1973 Endangered Species Act. In July, the National Academy of Sciences' committee on the Yellowstone grizzly, seated at the request of the Secretary of the Interior, released its report.

Cole had always said that there were two populations of grizzlies: the dump and campground bears the Craigheads had studied, and a separate population living in the backcountry on natural foods. With all those uncounted and unstudied bears out there, Cole estimated there were 350 grizzlies in the Yellowstone population. Echoing the Craigheads, the NAS characterized the belief in two populations as misleading. There was one population, and the NAS report said it had been essentially stable at 234 animals from the inception of the Craighead study until 1967. By 1973, Yellowstone's grizzlies had dwindled as low as 158, said the NAS panel.

Echoing the Craigheads' recommendation in their 1967 bear management report and in their presentations at the bear management meetings of 1969 to 1972, the NAS said that the basic unit of habitat for the Yellowstone population was an ecosystem containing not only national park land but also the surrounding national forests and state lands, and the bears needed to be managed that way.

Glen Cole's population estimates were essentially guesses, said the NAS report, and the park's research program since 1970 had "been inadequate to provide the data essential for devising sound management policies for the grizzly bears of the Yellowstone ecosystem." The report said it was not possible to know how the grizzly population was doing "without re-establishing a recognizably marked element." Further, the NAS stated that encouraging independent research like the Craigheads' was essential. Finding the Interagency Grizzly Bear Study Team organized the previous year to be ruled and controlled by the Park Service, the panel recommended that it be reorganized to include independent scientists.

Considering the fact that the NAS committee was composed largely of Starker's former graduate students and headed by a close colleague of his, Canadian wildlife ecologist Ian McTaggart Cowan; and that John Craighead resisted sharing his data with the NAS out of fear the committee would publish it before he did; the NAS report was an almost complete vindication for the Craigheads.

31

THE VERDICT

JENNY WALKER SAT in the witness stand with her long blond hair falling down past her shoulders and her hands folded in front of her. She had her father's gentle disposition and her mother's fortitude. Stephen Zetterberg approached the stand and handed Jenny a snapshot of Harry's room in the Walker farmhouse. It was a small room, furnished with a dark, antique wooden bed, an armoire, and a gun rack, on which hung a rifle, a fishing rod or two, and a pool cue. The bed was neatly made, the furniture gleamed with polish, and an American flag hung on one wall. Louise had turned the room into a shrine.

Zetterberg asked Jenny whether the room was Harry's.

"Yes, the flag on the back of his wall was what the National Guard—"

Jenny paused to compose herself.

"—When they gave him a funeral."

In her deposition the previous spring, Jenny had said that the realization that Harry was gone weakened everybody in the family, mentally as well as physically. "Especially my dad," she added.

Wallace had been holding his aching body together just long enough for Harry to come on full-time at the farm. After Harry's death, Jenny had seen Wallace's knee go out as he was jumping off a hay wagon, and she had watched him crumple to the ground. She'd seen him doubled over with back pain, limping out to grind feed corn for the cows.

Jenny had studied ballet for eleven years. She wrote poetry, had won an essay contest in high school, and thought she might like to be a journalist. It seemed that every dairy in Choccolocco passed from father to son: the Clark & Son Dairy, Hutto & Son, Carter and his son. But now there was no son. Jenny's sisters were married and had families of their own, and she was the last one at home. So when she graduated from high school, she didn't get married or leave for college. She quit going to ballet and helped her father and mother on the farm.

"How old are you?" Zetterberg asked Jenny.

"I'm twenty," she responded.

"And do you have any plans now—any foreseeable plans for leaving the farm?" asked the attorney.

"Not at the moment," Jenny answered. "I am not going to leave Daddy until I make sure that he is secure in what he is doing, because he needs me right now. If I left right now, he'd just have to close the dairy."

IN A PAPER he presented at a meeting of the American Association for the Advancement of Science in Philadelphia on December 29, 1971, Glen Cole had announced a milestone in what he characterized as Yellowstone's successful program to restore a natural population of grizzlies: no human being had been injured by a grizzly that year in Yellowstone. Starker Leopold later testified that the zero-injury rate contributed to his sense that the risky period of dump closure was over. At the trial, William Spivak had entered Cole's 1971 zero-injury report as an exhibit marked PP-2, to demonstrate that the Park Service had no reason—as in *Claypool*—to warn Harry Walker of unusual danger in June of 1972, when he entered the park.

Zetterberg meant to demonstrate otherwise. As he saw it, in an attempt to please Starker and justify his program, Cole had lulled the professor into complacency by not giving Starker all of the facts. Perhaps from what the Craigheads told him, or perhaps from information Martha Shell would later claim she and her husband obtained by breaking into a Park Service office and photographing rangers' logs of bear incidents, Zetterberg knew that records from Yellowstone's Lake District, which became ground zero for incursions by bears from Trout Creek dump when it closed, would give an entirely different impression from Cole's zero-injury rate in exhibit PP-2.

Zetterberg subpoenaed the logs for 1966 through 1971—handwritten ledgers of property damage, threats, injuries, relocations, and killings, kept by the rangers in each district. These he gave to Frank Craighead to study, and on the last day before closing arguments he put Frank on the stand as a rebuttal witness.

"Dr. Craighead," said Zetterberg, "have you spent some time studying this bear log and making notes?"

"Yes," replied Frank. "I have been reading over it and making notes. I found it very interesting."

"Can you tell the court what your summary shows?" asked Zetterberg.

"What they show is the number of incidents and actions, as close as I could tabulate them, for the various years," Craighead responded. "In 1966, there were thirty-three. In 1967, I tallied nine, which is unusually low. In 1968, with the closing of the Trout Creek dump, or largely closing it, there were eighty-four. In 1969, there were fifty-six. In 1970, there were fifty. In 1971, one hundred and one."

"Well, I see what they did here," said the judge. "The Park Service in PP-2 covered known and probable injuries from grizzlies. So it shows, in 1971, zero. Yet for the Lake District alone you find 101 bear incidents."

Hauk got it. Yellowstone was a powder keg, and something bad was going to happen.

Before closing arguments, the judge remarked that one thing he didn't understand was Superintendent Anderson's order to remove all radio collars and ear markings from Yellowstone grizzlies to beautify Yellowstone for the Centennial. It seemed to Hauk that given the danger inherent in the dump closure, the opposite should have been done. As a former naval intelligence officer, it made sense to him to employ whatever technology, like the radio collar, was available to keep track of grizzlies after they were caught hanging around places where people were present. Such a risk management process with the grizzlies could have been handled in the way he'd done it in the Navy, marking the position of enemy ships and aircraft in grease pencil on a sheet of glass, "So the bigshots could read it from the other side," he said. Certainly radio collars would cost something. How much? he asked Frank. Frank thought it might be about $3,000 per animal. Hauk said that whatever the cost was, he suspected that radio-collaring grizzlies might be considerably cheaper than paying off bereaved families of bear attack victims in the courts.

ON FEBRUARY 24, 1975, Jenny, Wallace, and Louise Walker listened raptly in the audience as Stephen Zetterberg presented his closing argument. The little noises in the courtroom—the shuffling of feet, the creak of a chair, the slight hiss of the air conditioning, the muted thump of the leather-upholstered doors when a casual onlooker or an attorney came in or out—all became noticeable. To Zetterberg, Spivak, and the judge it was another day at work, but for the Walkers it was an epic struggle between good and evil. Spivak—by all accounts a gentle, friendly man in private life—terrified Jenny Walker. With his shiny, bald head; his dour, stern presence; and the way he stared fixedly through his thick-framed glasses right past you, between questions; he had scared her half to death during the deposition at the courthouse in Anniston.

Addressing the judge, Zetterberg said that all that was known about Harry's character added up not to a scofflaw running from military service, but to a dutiful and hardworking young man who just needed a break.

"A last fling, to go out on a trip, one trip to Yellowstone Park," said Zetterberg, earnestly. "That boy had worked since he was ten years old!"

Harry would have gone home to stay on the farm, argued Zetterberg. He reminded the judge about Vikki Schlicht's statement to an investigating ranger that Harry had told her how much he liked working with the animals on the farm, and how he wanted to take her home with him, to meet his family and his horse, Comanche.

But as had been the case with William Claypool, Harry had not been provided with the information that something unusually dangerous was going on in the park, and thus he was not afforded the opportunity to make an informed decision about whether it was wise to camp out. The Park Service had sort of blundered into the dump shutdown with the reduction of trash at Trout Creek, argued Zetterberg, and the precipitous rise in bear incidents of 1968 and the grizzly attacks of the following year had increased the rush to judgment at the bear management meeting of 1969.

Zetterberg reminded the court that Starker's testimony—that it was anyone's guess what would happen when the dumps were closed—had no longer been true in October of 1969, when dump closure was approved. By then the Park Service had seen enough mayhem just from trash separation at Trout Creek to know there was going to be more trouble. Yet the agency didn't warn visitors, just as it had failed to do in *Claypool*. And although Starker Leopold had recommended continued tracking of grizzlies during dump closure, not only did Anderson and Cole forbid the Craigheads to put additional collars on bears, but they started cutting them off grizzlies that already had them. Then, Cole had misrepresented the tenseness of the dump closure situation to Starker. Sure, there hadn't been a grizzly attack in 1971, but the 101 bear incidents at Lake alone would have told Starker that there was about to be one.

Then there was the problem with signs.

A sign seemed like such a simple thing—could it really be true that a placard with just a few words of warning about bears might have made the difference for Harry between life and death? The father of Andy Hecht, the nine-year-old boy whose remains Jim Brady had strained out of a hot pool in June of 1970, mounted a multiyear campaign to force the Park Service to be more proactive in marking hazards. But the service liked to err on the side of not walling people off from nature with protective barriers and scary signs.

Zetterberg pointed to an enlargement of the photo he'd taken at the Roes Creek trailhead five years after Smitty Parratt's mauling. "I would like Your Honor to look at the Glacier sign there," he said. "That is the sign that was in Glacier Park, and Cole knew about it. He knew about the sign." Yet Cole had not seen fit to post similar signs at the beginning of the boardwalks on Geyser Hill, said Zetterberg.

Sure, there was a sign at the North Entrance station, where Harry, Vikki, and Phillip entered Yellowstone, to the effect that the park's bears were wild and potentially dangerous animals. And there were signs and brochures elsewhere warning people not to feed or molest them. The problem was, argued Zetterberg, the overall message the Park Service had given the public was that if you left the bears alone and didn't feed them, they would leave you alone. Both Vikki and Phillip had said in their depositions that they and Harry agreed with this rule and felt disdain for the tourists they saw feeding a bear on their way to Old Faithful. But while this rule might keep you out of trouble with black bears, it wouldn't keep you safe from grizzlies after dump closures, said Zetterberg.

"The point is this," said Zetterberg. "The Government was taking a risk by closing the dumps and sending bears hungry into the park." But the risk was not borne by the government, he argued. It was defrayed onto people who didn't know about it: Harry Walker and the unknown camper at Firehole River.

Zetterberg said with all that he had done in over two years of work on the Walker case, he was haunted by things he might have said to better represent his clients, whom he believed genuinely deserved the court's help.

"I hope Your Honor will consider the evidence . . . I hope Your Honor will grant this family a substantial award, and I hope Your Honor will send a message to the Interior Department via a substantial award so that this kind of thing will never happen again," concluded Zetterberg, and sat down.

"All right. Let's hear from the Government," said Judge Hauk.

IN HIS CLOSING argument, William Spivak pointed to Harry Walker's status as an AWOL national guardsman as not only a flaw in character but a potentially fatal one. A brand-new visitor's center stood right by the hotels and the Old Faithful Geyser, and there was a big ceremony to dedicate it while Harry and Phillip were there. During the summer it was open in the evening, and sometimes all night. Why, then, had Harry and Phillip not gone inside to inquire about camping permits, which would have provided them with information on bears and proper food storage, asked Spivak. Maybe because Harry did not want contact with authority, he suggested. He referred the judge to an exchange between himself and Phillip Bradberry, during the young man's deposition in Anniston:

"Was Harry running?" Spivak had asked.

"No—" Phillip dissembled, then admitted, "Yeah, yeah. He was AWOL from the guard."

Maybe that was why Harry had avoided talking to the rangers about where to camp, said Spivak. He was on the run from the law.

When Jim Brady's rangers located Harry's body and the nearby camp, they found food stored on the ground next to the tent, said Spivak, some of which seemed to have been spread around by the bear that killed Harry. Had Phillip and Harry bothered to get camping information, they would have been told to hang their food in a tree, he said. And there was more than enough signage, pamphlets, and uniformed employees to inform people about the danger of bears.

Further, Spivak challenged Frank Craighead's testimony connecting the possible human scalp and torn-up camp at the Firehole River and bear 1792, which Jim Brady captured and relocated, and which later killed Harry Walker. Frank Craighead had asserted the connection based on the notion that the cans at the abandoned camp had been mashed, but not punctured, by a toothless bear. In support of this connection, Zetterberg had entered into evidence cans with the same telltale signature from Harry Walker's camp. But Spivak pointed out that Frank's claim about the Firehole cans was based on hearsay from the young ranger he'd met in West Yellowstone. In fact, Spivak noted, Ranger Tom Cherry, who had been sent to investigate the Firehole camp, had testified in a deposition in 1973 that the cans he saw there had the punctures typical of a bear that had all its teeth. Therefore, bear 1792 was not the bear that tore up the Firehole camp. Spivak's intent in making this argument was to disconnect from the rest of the case any liability derived from Jim Brady's decision to spare the bear and relocate it, in 1970.

Then Spivak made an even broader denial of government responsibility: aside from all the specifics of the case he had been arguing thus far, the federal government had a general immunity from this type of lawsuit under a doctrine known as the "discretionary function exemption" in the Federal Tort Claims Act.

The act's congressional framers had included in it an exception, a "discretionary function clause," which held the government blameless in decisions that "required reasoning and judgment" about whether or not to take actions that were optional and not imperative under the law. This rather vague principle was apparently included in the act to keep the business of government from grinding to a halt over fears that every discretionary function—whether to paint postal delivery trucks bright orange for traffic safety, or keep them red, white, and blue—would be endlessly second-guessed and result in frivolous lawsuits. Such was the nature of dump closure, said Spivak. It was within the discretion of the Park Service to do it, the service took its best shot, and whether or not it came out perfectly, the agency was immune to prosecution for it.

Spivak rested his case.

ANDY HAUK HAD been appointed to the federal bench by President Lyndon B. Johnson, and when the appointment was announced, Johnson actually telephoned Hauk to congratulate him. According to Hauk, The president offered a piece of advice in his Texas drawl: "Rule quickly and decisively, and sleep like a baby."

Hauk took the president's advice. It was Hauk's practice to rule from the bench as soon as the attorneys completed their arguments.

"All right," he announced when Spivak and Zetterberg had finished, "I will never know any more about this case than I do right now. I am ready to rule."

He ruled in favor of the Walkers.

Each attorney in the case had filed a sort of wish list of what were called "findings of fact" for Hauk to use as a cheat sheet in constructing his ruling. Hauk proceeded to make a few edits to Zetterberg's and then began adding up damages, reciting his figures to the court reporter as he did the arithmetic on a piece of paper in front of him. The figures had been provided by a consulting economist employed by Zetterberg, who calculated the life value of Harry Walker's work on his family's farm, and who, back in 1965, had calculated Smitty Parratt's medical bills, the cost of his rehabilitation, and his vocational training all the way through college.

"Let's see what that adds up to," Hauk said when he finished. "That is $87,417.67."

In the audience, Wallace, Louise, and Jenny hugged one another and wiped their eyes. That was a lot of money in 1975. Starker Leopold's round-trip airline ticket from Oakland to testify in Los Angeles had cost the US attorney forty-one dollars. The money wouldn't bring Harry back, but it would help keep the farm alive.

THE STORY, PICKED up by the wire services and printed in newspapers all over the United States, stated that Judge Hauk had found that the federal government "willfully and intentionally failed to post signs in camp grounds and developed areas to warn persons of the danger from bears." It went on: "Yellowstone officials should have continued its monitoring of grizzlies—and could have for little expense. The monitoring, by radio, had been ended shortly before . . ."

In Missoula, John Craighead gave his secretary a letter to type up and a stack of addresses to send it to. Calling attention to the federal court decision, the duplicate letter basically said: "We've been vindicated." The brothers had lost their fight to remain in Yellowstone, but the Park Service had egg on its face again about its management of grizzlies. It sounded like the agency hadn't learned much in the eight years since the night of the grizzlies. The bad press kept things moving in the direction they were already going, out of the park and into the Greater Yellowstone Ecosystem.

Five months after the Walker verdict, on July 28, 1975, the US Fish and Wildlife Service listed the grizzly bear as a threatened species in the Lower Forty-eight contiguous United States. The Endangered Species Act made it illegal for any agency to do something that would negatively affect a threatened animal. Agencies were required to determine whether any project—logging, the construction of ski lifts, oil and gas exploration—would affect the species. In order for agency biologists to comply, they needed to map a threatened animal's habitat to know whether a given project fell within it. The effect was to shift attention from merely protecting animals to protecting animals as well as their habitats. Another effect was to keep interest alive in building new populations of grizzlies, by transplanting them to places the species had once occupied.

In 1971, while working with Joel Varney on computer modeling of the Yellowstone grizzly population, the Craigheads also got interested in mapping grizzly habitat from aerial photographs and, later, from satellites. In one outgrowth of the discussions about moving Yellowstone bears farther away from trouble, the brothers suggested that it would be interesting to look at Montana's Scapegoat Wilderness, which no longer had grizzlies, as a place to put some. The brothers' proposal was funded by the Forest Service, and in 1972 and 1973 they worked back and forth between identifying on aerial photographs the signatures of things grizzlies ate—berry patches, whitebark pine groves, and meadows full of wildflower bulbs—and backpacking into the wilderness to see if they were right.

Meanwhile, the Craigheads had been working since 1969 on another idea that was three decades ahead of its time: satellite tracking collars. The project involved conversion of navigation buoy beacons to track elk and grizzlies from NASA's Nimbus satellites. The effort was limited by the technology of its time. A device the brothers tested on an elk in 1970 was as big as a horse collar and weighed twenty-three pounds. But their research presaged the GPS tracking systems that are now ubiquitous in wildlife research. The Craigheads were visionaries. In one of Frank's spiral-bound pocket notebooks he drew a sketch of a solar-powered tracking device that could be mounted on the lower back of a songbird, powered by miniscule solar panels affixed to individual tail feathers. Nothing close to that level of miniaturization existed at the time. Today biologists put radio tracking devices on butterflies.

AT YOSEMITE IN 1975, Jim Brady's rangers enforced park regulations requiring visitors to store their food in their cars, but at night the bears broke vehicle windows or tore open doors and got to it anyway. In the backcountry, campers were told to hang their food in trees, but black bears were good climbers, especially the cubs. Out on foot at night with his flashlight and dart rifle, David Graber watched female black bears teach their cubs—who are adapted to scurry up trees for protection—to climb up and claw down backpackers' food bags, from the branches of pines.

That spring, David Graber expanded his study east up the Tioga Pass Road to White Wolf campground, Tenaya Lake, and Tuolumne Meadows; north to Hetch Hetchy; and up the foot trails to the hanging canyon of Little Yosemite Valley. It was a really bad year for bear problems. At White Wolf, he watched two bears bash their way into a car and fight each other inside, tearing apart the interior. Just before Memorial Day, a man yelled at a bear he found eating his food, which he had left unprotected in a campground. The bear smacked him with its paw, causing abrasions. On June 17, a ten-year-old girl was feeding fresh cherries to a bear near the Yosemite Lodge. When the cherries were gone, the bear roughed her up. Five days later another child approached a bear from behind, thinking it was a pet. The bear inflicted lacerations with its claws. On June 24, a bear tore through the thick canvas wall of a rented tent cabin and swatted the startled occupant, causing abrasions.

On July 5 at another tent cabin, a black bear was eating the contents of an ice chest left outside. The noise awakened an occupant, who looked out to see what was happening and was clawed by the bear. On July 10, a bear stepped on the face of a man sleeping on the ground, causing minor injuries to his cheek. Two days later another man sleeping out without a tent was attacked by a bear. His lacerated scalp required seven or eight sutures. On July 17, a man sleeping in a tent heard noises outside and opened the tent flap, at which point a bear attacked him, causing minor injuries. Yet another man in a sleeping bag was either attacked or stepped on by a bear but was uninjured. Another attack on July 30 left a victim with scratches and bruises.

In a backcountry campsite the night of August 2, a man in a sleeping bag was awakened by a bear tearing into his backpack. The bear inflicted lacerations to the man's head. On the night of August 13, a bear entered another backpacker's camp, where it pulled down food sacks hung in a tree, ate some of the contents, and then retreated to the edge of the woods. The camper tried to salvage what was left of his groceries. The bear charged and clawed him, cutting one of his hands and flaying open one of his forearms. Alone and bleeding, the man limped down the trail to find help. It was a cold night, and by the time he staggered into the summer ranger station in Little Yosemite Valley he was shivering violently and slipping into shock. The rangers gave him first aid and evacuated him. The following day, a bear knocked a man down in a campground and inflicted twelve-to-fifteen-inch-long lacerations on his back. On September 4, another man in a sleeping bag was awakened by a bear standing on his legs. When he attempted to wiggle out from under it the bear clawed his face, cutting him under one eye. In addition to all of these injuries, by the end of the season Yosemite rangers had recorded 979 incidents of property damage, totaling $100,000 in 1975 dollars.

JAN VAN WAGTENDONK, Yosemite's research scientist, who shared responsibility with Starker Leopold for supervising David Graber, summoned Graber to his office.

"Dave, you're going to finish what you need for your master's, and we're not going to have enough data to really change the management of bears, are we?" said Van Wagtendonk.

"No, probably not," Graber replied.

"I have a proposition for you," said Van Wagtendonk. "Do you like being here and working on bears?"

"Yeah, it's the happiest time in my life," replied Graber.

"Yes, I thought that it might be," said Van Wagtendonk. "I have a suggestion. If you leave, I'm going to have to hire another kid to do the next two years, or I'll have to start from the beginning. It would be more profitable if I kept you. How would you think about doing a PhD on black bears and staying here?"

"Well, you know, that sounds like a lot of work," Graber mused. He'd never considered going farther than a master's degree, which could get you a job as a field wildlife biologist.

"Aren't you having a whale of a time?" Van Wagtendonk pressed Graber.

"Yes, I'm having the best time I ever had in my life," Graber admitted.

"It'll continue," concluded Van Wagtendonk.

"Okay," Graber answered, "but I have to get accepted to the PhD program at Berkeley, and I have to talk with my wife about whether she can live with that."

Van Wagtendonk nodded. "It's been taken care of. Starker and I have already talked to your department. And your wife."

Graber remembered thinking, "Hey, don't I get a say in this thing?"

IN THE FORTY years he would spend as a scientist in national parks, David Graber liked to say that he had invented only two things. The first was mail-order gourmet coffee, ordered by credit card. Staying up all night chasing bears, he was so tired on some days that he was hallucinating. At one point he left a woodstove burning in his canvas-walled tent cabin at Tuolumne Meadows and burned the place down, with all of his gear. He and his research assistants became addicted to the extremely strong Peet's Coffee they bought in Berkeley. One day it became clear to Graber and his assistant that they didn't have enough coffee to make it until their next resupply. It was a serious threat to their research. Graber called up Peet's original North Berkeley store and asked the manager to send him a box of coffee. A negotiation followed. Peet's didn't ship retail coffee. Graber begged, a test case was made, and from that time forward, fragrant packages were periodically delivered to Yosemite's administration building, and Peet's began shipping to other customers.

Graber's second invention was the type of bear-proof food storage box that is now everywhere in government campgrounds. In the autumn of 1974, a staff meeting was convened in the auditorium of the visitor's center in Yosemite Valley to discuss what had been learned about the bear problem during that difficult summer. At that point Graber had not read George Wright's suggestion about secure food storage, or Jim Reid's from 1969. But he'd seen a heavy steel locker used to store the rangers' groceries at the wall tent that served as a summer backcountry ranger station in Little Yosemite Valley, and he'd watched with considerable interest as bears spent hours trying—and failing—to get into it. Graber spoke up at the meeting and suggested that something like that could be installed at each campsite in Yosemite. But Yosemite's superintendent and chief ranger scoffed at the young graduate student's idea, saying it would be vastly expensive.

George Wright must have encountered a similar argument within the agency in the early 1930s, because, without giving credit to the source of this penny-wise, pound-foolish notion, he refuted it in his 1933 report: "When we say we cannot afford a thing, we mean that we do not value it as highly as we do something else. But either the property damage done by bears is worth doing something about or it is not worth considering."

Rick Smith, a widely respected ranger who worked under Jim Brady, was in the room when Graber made his suggestion. Smith had a master's degree in English literature and was known for his considerable powers of verbal persuasion. Smith could talk a belligerent drunk into the back of a patrol car without resorting to force. Now he spoke up to defend Graber's idea. Serious discussions followed. Graber was asked to suggest where a pilot installation could be done. Wisely, instead of picking the largest and most complicated problem in the valley, he suggested the smaller White Wolf campground on the road to Tuolumne Meadows. Installation began in 1977, and with enforcement—people had to use the lockers in order for them to work—White Wolf's bear problems were solved. After a half-century struggle with food-habituated bears, it was the first such installation anywhere.

The collective momentum of the night of the grizzlies, the Walker trial, the listing of the grizzly under the Endangered Species Act, Galen Rowell's publicity about Yosemite's bear graveyard, and the obscene rate of property damage and injuries at Yosemite made it possible for one young protégé of Starker Leopold to do something that would, over time, bring to a close the era of large numbers of food-conditioned bears in national parks.

32

THE APPEAL

THE WALKERS' VICTORY in federal court was short-lived. The US Attorney's Office submitted the case to the Ninth Circuit Court of Appeals, where a three-judge panel was appointed to review it. According to the lawyer who managed the case for the Department of the Interior, two of the three had ruled against the victims in previous bear mauling cases—a pretty exotic category in a world of civil rights violations, interstate commerce infractions, and imminent domain condemnations for freeway construction. It's unclear whether this was arranged or an unfortunate matter of chance, but on December 3, 1976, the appellate court reversed Judge Hauk's decision. Because the case originated in Wyoming, under federal law, Wyoming statutes applied. In Wyoming if a plaintiff contributed at all—even 10 percent—to their own injury or death, they or their estate could not collect anything. The appellate court found that Harry and Phillip had contributed to the incident by camping illegally and leaving their food on the ground instead of hanging it in a tree. Further, the appellate judges ruled that, as Spivak had argued, the discretionary function exception to the Tort Claims Act applied in this case. The Park Service had the discretionary authority to close the dumps and was protected from liability stemming from the method it chose to carry out the job. In 1977, Stephen Zetterberg appealed the case to the US Supreme Court. On June 20, 1977, the Supreme Court declined to hear the Walker case and let the appellate ruling stand.

Zetterberg was not one to quit. He believed in representing his clients until every possible gambit had been exhausted. He prevailed on Alabama Democratic senator John Sparkman to organize the passage of House and Senate relief bills to return to the Walkers the money Judge Hauk had awarded them, plus about $15,000 in legal fees for Zetterberg & Zetterberg, which had worked for over four years on the case and had not been paid. In a letter explaining this move to the Walkers, Zetterberg enclosed a clip of a newspaper story about a federal audit that revealed that the House of Representatives had spent $70,000 in a single year on liquor. If the government could afford that, wrote Zetterberg, it could afford to help the Walkers.

Senator Sparkman introduced the Walkers' relief bill in 1978. That spring, Gerry Tays, a former Yellowstone law enforcement ranger who had worked under Jim Brady on the search for Harry Walker's body, happened to read an article in the *Washington Post* about the gifts bestowed on Americans in congressional relief bills. The article contained a passage about the Walkers' bill.

By some terrible mechanism of fate that seems to have prevailed in the Walker case, Tays was the very horse patrol ranger whom Jim Brady had assigned to ride around Old Faithful and root out illegal campers on summer nights in the early seventies. He was now working in the Park Service's Washington, DC, offices on Park Service legislation in Congress.

Tays was incensed. In his opinion, Harry and Phillip had been the victims of their own decision to camp illegally and to not hang their food in a tree. Tays got out his *Congressional Staff Directory* and telephoned an aide to Malcolm Wallop, the arch-conservative freshman senator from Wyoming. In another stroke of fate, the senator's aide had once worked as a seasonal ranger at Grand Teton National Park and was familiar with the Walker case. He agreed that private relief bills like the Walkers' were the sort of socialist, big-government magnanimity that Senator Wallop's constituents had sent him to Congress to stop. The Walkers' bill had been approved by the Senate Judiciary Committee and placed on the unanimous consent calendar, where it would have breezed through without much trouble. But notified by his aide, Wallop objected to it there, and it was set for debate on the Senate floor. The elderly Senator Sparkman had arranged for the junior senator from Alabama, Democrat James Allen, to shepherd the bill through the Senate. On June 1, 1978, the day before Allen was supposed to argue the Walker relief bill on the Senate floor, he died of a heart attack on the last day of a vacation on Alabama's Gulf Coast. After Allen's death the bill lost its way and was never passed.

EACH MORNING IN the mid-seventies, Starker Leopold would sit down at his oak desk in his subterranean office, surrounded by his father's books, his students' doctoral theses, and his mounted game heads on the walls, and write eight to twelve letters in longhand on ruled paper, which his secretary would then type on university letterhead backed by carbon paper and yellow onionskin, which would become the office's file copy.

The range of his correspondence was remarkable. Government officials and scientists from all over the world sought his advice, yet he dutifully answered a letter from a child in Michigan asking for information on the extinction of the passenger pigeon for a school report. A housewife in Massapequa, New York, wrote Starker asking for advice on the culture of edible snails. A former student wrote from Mexico asking if the professor could help locate a copy of his now out-of-print book on the wildlife of that country. Starker wrote back with the addresses of used-book stores in Mexico City. Senator Ernest Gruening of Alaska requested Starker's analysis of the US Plywood–Champion Paper Company's plans to log the Tongass National Forest. Starker answered.

A wildlife biology undergraduate at Colorado State wrote Starker asking six highly detailed questions about the habits and management of sea otters. The letter betrayed a failure to conduct basic library research on the subject prior to asking the famous professor to do it for her. Starker answered, patiently citing available readings. A woman from the Audubon Society in Morro Bay, California, wrote him regarding the status of species of migratory geese. Starker answered the letter with a seemingly encyclopedic knowledge of the relationship between the numbers of such geese returning to winter in central California and the number of arctic foxes and the timing of the spring breakup of ice in Alaska, where the birds raised their chicks.

An editor at Macmillan Publishing Company wrote asking Starker to pen a textbook in wildlife management. Dr. Michael G. Ridpath of the Division of Wildlife Research at F. C. Pye Wildlife Research Lab in Darwin, Australia, wrote Starker asking whether he had heard any accounts of eagles carrying off schoolchildren. Starker replied, citing three such incidents, one in Norway in which an infant in swaddling clothes was snatched away and carried to the eagle's treetop nest, another in Canada in which a crippled child was carried away, and one more involving a wedge-tailed eagle in the requesting scientist's own country, which Starker said he assumed the Australian scientist already knew about.

Starker Leopold was a sort of human search engine, before search engines, on wildlife and the environment. But more than that, his sober yet immensely energetic and friendly manner brought people of all walks of life to trust his judgment. With his office a half story belowground, its window affording an ant's-eye view of the undersides of the ornamental plantings against the building's exterior wall, it was as if Starker Leopold's immensely authoritative voice issued from the very soil; as if he spoke for the earth itself. With the growth and diversification of the ecological sciences, there is certainly no one person living today who has been given such broad authority in those fields and is consulted by and relied upon by so many. And yet Starker was only human. In 1971, at the height of the struggle with the Craigheads, he had written a history of the radio tracking of mammals in a proposal for telemetry research that he wanted to do in the Sierra Nevada. In it, he cited over a dozen scientists who had developed the new techniques, but he omitted any mention of the work of the originator of the first practical radio collar, Frank Craighead.

Starker never stopped believing that the judgments that would be needed to properly manage national parks in an era of extinctions would require more scientists, and he kept cracking the whip over the Park Service. Having resigned as chairman of the Natural Sciences Advisory Committee in 1972, he was immediately appointed to yet another body, the National Parks Advisory Board. In that capacity, in 1977, a new Park Service director asked Starker and a fellow member, Durward Allen, to assess what progress had been made in bringing scientific management to the national parks.

"We find no major fault with the science program," Starker and Allen wrote in their report to the director in July, "except that it is inadequate to meet its implied responsibilities." At the time, research scientists constituted just over 1 percent of the agency's staff. Over the next decade their number shrank by three, to one hundred for the whole country.

Against this gloomy picture, there were a few bright spots. One was Yellowstone, which, as it continued to generate controversy, also continued to build its staff of PhDs. Another was Everglades, where under the regional influence of the Tall Timbers group an experimental prescribed burning program began in 1958, even before the first western forest was burned at Sequoia and Kings Canyon. In the 1970s, Everglades' enlightened superintendent, Boyd Evison, established a research center there. Another center of science was Sequoia and Kings Canyon, where Evison transferred in 1978 to succeed John McLaughlin. Evison promptly gave up the generously sized superintendent's residence at Ash Mountain for conversion into a research center, which soon became the office of, among others, Dr. David Graber.

Starker Leopold retired from UC Berkeley in 1978 but continued to work as a professor emeritus. In 1979, the last of the open dumps around Yellowstone was closed at Cooke City, on the park's northeast side. That year David Graber defended his thesis, "Ecology of the Black Bear at Yosemite," and became the PhD he had never planned on being before Jan Van Wagtendonk asked him to stay on in Yosemite for another two years. Starker arranged an interview for a teaching job at Yale, which Graber attended halfheartedly, but on his return he confessed to Starker that his heart was in the national parks. In 1979, with Starker's blessing, Graber was hired at Sequoia and Kings Canyon by Bruce Kilgore's replacement, Stanford botanist Dr. David Parsons, to continue black bear ecology studies and to work with the California Department of Fish and Game, the Fish and Wildlife Service, and the Forest Service on what became a highly manipulative, multidecade program to restore herds of critically endangered bighorn sheep in the Sierra.

SEQUOIA AND KINGS Canyon had the fourth-worst bear problem in the National Park Service, after Glacier, Yellowstone, and Yosemite. In the spring of 1980, when Graber started work there, not one of the parks' twelve hundred front-country campsites was equipped with a food storage locker. The regulation requiring visitors to store food and ice chests in their cars was as futile and destructive of visitors' property at Sequoia and Kings Canyon as it was in Yosemite. The year Graber arrived, the parks recorded 467 bear incidents, most of them car break-ins and raids on backcountry campsites. Injuries were uncommon, but hair-raising bluff charges and other scary situations involving emboldened, food-conditioned bears weren't. In Giant Forest a bear broke a window and climbed into a rental cabin at night, landing on a bed in which a young woman was sleeping. Her screams awakened members of her family in an adjoining room, who charged in to save her. In the melee that followed, the young woman was injured and the bear fled.

Around the same time Graber arrived, Sequoia hired a new resource management biologist, Harold Werner, whose job was to carry out the recommendations made by scientists like Graber. Werner had flown helicopters in combat in Vietnam and, when he got out, studied wildlife biology. He got a job at Everglades, where he banded sooty shearwaters on the Dry Tortugas and met Frank Craighead's father, Frank Sr., who had retired from his job in Washington and was also doing bird research there.

Werner was an exceptionally gentle man. In the fall of 1980, he had to kill his first recidivist bear at Sequoia. He knew the bear, having followed its career. It had been given a name, Willie Wonka, which the Park Service later forbid biologists to do because they got too attached to bears with names. Once Willie Wonka was trapped and anesthetized, Werner took out a .300 Savage rifle. His new wildlife technician, Mike Coffey, offered to do the job. Werner declined. He fired a single round into the sleeping bear's chest, cleared the chamber, set the rifle down, and wept. Werner became very interested in installing food lockers.

The parks already had bear-resistant garbage cans, but the rangers had problems getting the maintenance division to empty them on busy summer nights, so visitors would pile overflow garbage against them and the bears would get to the food anyway. With growing numbers of visitors and decades of failure to prevent food conditioning in bears, even as food lockers went in, in a single week that August, bears broke into twenty-one cars at Lodgepole campground. Nineteen-eighty-two became the worst year for bear incidents in Sequoia and Kings Canyon's history thus far, with 621 incidents of property damage totaling $67,000 in costs, physical threats such as bluff charges, and a handful of injuries to park visitors.

Graber instituted a highly organized system of recordkeeping. Bears that were trapped in campgrounds and other developed areas became, as was true of the Craighead study, part of the research population. They were each marked with a metal ear tag to which was attached a fabric flag bearing a distinctive pattern of geometric shapes in bright, contrasting colors, so they could be distinguished at a distance. The number on the tag was also tattooed on the inner surface of the bears' lower lips so they could be identified if their tags fell off. Some were fitted with radio collars. A dossier was opened on each marked bear.

When a bear had been relocated repeatedly and persisted in being a miscreant, a tribunal known as the Bear Management Committee assembled at Park Headquarters at Ash Mountain to decide its fate. The committee included Graber; Harold Werner; Werner's boss, the chief of resource management; and his wildlife technician, Mike Coffey. When the decision was thumbs-down, it generally fell to Coffey to carry out the sentence, not because Werner shirked that duty, but because he was tremendously busy with other work.

Sometimes killing a bear went easy, and sometimes it didn't. Deep in the granite canyon of the Middle Fork of the Kings River, bear Number 905 had been seen breaking into cars at the Sentinel Rock parking area. The Bear Management Committee condemned it to death, and Mike Coffey drove into the canyon to stake out the parking area at night, bringing a twelve-gauge shotgun, because there were often backpackers on the trails in the Middle Fork, and a rifle bullet would fly too far. The bear appeared, like the habitual car burglar it had become, after midnight. When Coffey approached, it climbed a tree. Coffey fired a rifled slug at it, hitting the bear in the neck. A second shot hit it in the shoulder. The bear did not fall. Coffey fired again, hitting the bear in the left buttocks, near the tail. The bear fell out of the tree, got up, and fled. Coffey gave chase. He followed the blood trail to the banks of the Kings River and waded the icy water, alone in the dark. On the far side he lost the blood trail up a mountainside. He climbed back down and went to get what sleep he could.

In the morning he returned, again waded the Kings River and climbed the mountainside where he had lost the trail the night before. At 12:30 that afternoon, Coffey found the bear alive and bleeding up in the timber. He shot it for a fifth time, in the head. A night or two later he was seen disconsolately nursing a beer at a bar in Three Rivers, just outside the parks. No one liked doing these things, and, like Harold Werner, Coffey became intensely interested in lockers and any other measures that might keep bears away from human food and, ultimately, save their lives.

BACK IN APRIL of 1975, two months after the verdict in *Martin v. United States,* Starker Leopold had addressed a meeting of park superintendents in San Francisco on the subject of managing and restoring park ecosystems, which the Park Service referred to as "resource management." Looking back twelve years on his Leopold Report, Starker said there was one sentence he would like to have changed in it: "As a primary goal we would recommend that the biotic associations within each park be maintained, or where necessary, re-created, as nearly as possible in the condition that prevailed when the area was first visited by the white man."

"That was too narrow and restrictive, and the phraseology was bad," Starker told the superintendents. "In other words, it implied stopping the clock. And what we're talking about here is dynamic processes that don't stop. Biotic associations change, constantly, naturally."

What Starker was acknowledging was the decline of Clementsian succession. Ecosystems were no longer headed for a known endpoint. Indeed, they might have several potential stable states. When Doug Houston published his 1982 summus *The Northern Yellowstone Elk,* a professor from the University of British Columbia wrote in the foreword that elk might have one state when regulated by predators and another when regulated by vegetation, both of them valid. By 1982, nine years after Drury and Nisbet's 1973 paper, Doug Houston, Mary Meagher, David Graber, and David Parsons had abandoned the Leopold Report's focus on re-creating a certain *product*—sequoia groves or the Northern Range as Euro-Americans first saw them—in favor of supporting the natural *processes* that formed, and kept re-forming, them. If scientists supported processes like fire and unmanaged elk, the thinking went, they would turn out natural results.

At Sequoia and Kings Canyon, David Parsons and a young fire scientist in the park's Division of Resource Management, Tom Nichols, presided over the fourteenth year of prescribed burns in the giant redwoods. Nichols was an intellectual who looked more like a computer programmer than a park ranger. He had a shock of unruly, kinky dark hair, wire-rim glasses, and a tendency to toss off finely honed ideas in rapid-fire sentences with a shrug of his shoulders as if what he was saying was obvious to everyone. He would rise through the Park Service to become the agency's chief of fire and aviation. Although David Graber was a wildlife ecologist, David Parsons a botanist, and Tom Nichols a general-purpose fire scientist, with a small staff and lots of projects they tended to collaborate across the lines of their disciplines. Parsons and Graber were the senior scientists. Tom Nichols's job was to lay out controlled burns and write a prescription for the weather conditions and fuel moistures within which they could safely be kindled (thus the term "prescribed burn"). He got good enough at it that he wanted a little more direction from Parsons and Graber about exactly what kind of fires they wanted.

"I'd gotten to the point where I could do whatever we wanted with fire," he said later with a shrug. "You want a small fire? I'll give you one. A big, intense fire? You got it. I could make fire jump through hoops. We had been thinking that just putting fire back in would take care of it. But what kind of fire did we want?"

In the beginning, prescribed burning was thought of as a means to repair the ills of previous fire suppression and put the forest back in condition where lightning fires could maintain it. To this end, from the early sixties until the early seventies, facing huge fuel loads, Biswell, the San Jose State group, and Kilgore had done some fuels management before burning. In some cases they would have forestry crews cut, slash-pile, and burn dead limbs, brush, and white firs so they wouldn't burn down sequoias. After 1973, with the emphasis on process over product, mechanical fuel treatments were mostly discontinued at Sequoia and Kings Canyon in favor of pure prescribed burning, in which the process of fire was allowed to produce whatever results it wanted.

By the early eighties, however, the expectation that prescribed fire could be used to return the forest to a natural condition, after which lightning could be relied on to take over, failed in the face of evidence that before the parks existed, Indians had set at least as many fires as lightning, or a good many more. By 1982 it became clear that there would be no putting fire back in and walking away. There was another problem too: Tom Nichols could lay out a fire that would burn white fir saplings and encourage germination of sequoias, but he couldn't make one big enough to take out the larger white firs that grew up during fire suppression without burning down sequoias.

Parsons, Nichols, and Graber were not the only ones who saw this. In August of 1982, Tom Bonnicksen, a professor of forestry at the University of Wisconsin, and Edward Stone, who had been his faculty advisor UC Berkeley's School of Forestry and Natural Resources, published a paper in *Ecology* challenging the idea that prescribed burning would restore giant sequoia groves to their "natural," pre-fire-suppression condition. Although controlled burns had reduced fuel and stimulated growth of sequoia seedlings, Bonnicksen argued that they were not restoring a "natural" forest with the same ratio of sizes and species of trees that prevailed before Euro-Americans stopped fires. Instead, Park Service burning was perpetuating a new kind of age-stratified, manipulated forest, with few white fir seedlings and saplings but lots of moderate-to-large white firs. For a landscape in which to enumerate the failures of Sequoia's burn program, Bonnicksen chose Redwood Mountain, which was to Bruce Kilgore, Harold Biswell, and Starker what Kitty Hawk was to the Wright brothers.

Bonnicksen had entered UC Berkeley in 1968 as a forestry undergraduate and took classes from Starker and Biswell. In 1970, when he graduated, he went to work as a summer ranger naturalist at Sequoia and Kings Canyon, where he led walks on Redwood Mountain to show off the Park Service's prescribed fire program to visitors. In the course of these hikes he came upon the remains of white firs that had been cut down with chainsaws and their stumps blasted with dynamite. Sometime in the early seventies—concurrent with Cole's Natural Regulation—the mechanical removals stopped at Sequoia. It was as if everyone got some new religion of letting nature do its own work. Which it wouldn't, said Bonnicksen.

The only thing that could reproduce the structure of a presuppression sequoia forest, argued Bonnicksen, were loggers with careful plans for which trees to take out and which to leave. Such plans could be devised by doing age regression studies of aggregations of older trees, using the relationship between their present ages to turn back the hands of time and reveal what the forest had looked like before firefighters.

For Graber, Parsons, and Nichols, the idea that nature could not be trusted to restore itself without chainsaws flew in the face of the idea of wilderness as a refuge for natural processes. Bonnicksen charged Graber and Parsons with being anti-human. "I believe we belong here," he said of human civilization.

Graber grudgingly admits that Bonnicksen was right about some things. But he never liked Bonnicksen much. Maybe, in part, that was because Bonnicksen was associated with a painful period when Graber and his fellows lost a clear sense of what they were doing. They had started out from the Leopold Report headed for a product. Then they abandoned that to support a process. And then they realized that they would always be part of the process.

IN THE SPRING of 1983 at Sequoia and Kings Canyon the rain and snow never seemed to stop, and the Kaweah River rumbled, milky white with foam and yellowish with silt, through the oak woodlands in the steep, green foothills around park headquarters at Ash Mountain. By then it had been discovered that a bear had broken into the ranger station in the Mineral King Valley and lived inside it for a while, happily sucking the contents out of canned goods.

Late in March, Mike Coffey left park headquarters and drove into the San Joaquin Valley, where the air was heavy with the scent of orange blossoms, headed north toward San Francisco, where he had arranged to meet the bear keeper at the Fleishhacker Zoo. Harold Werner had been working with private suppliers and biologists in other parks to develop a device that no backpacker had yet seen but that today nearly every backpacker owns. It was a lightweight plastic food canister made out of a piece of eight-inch black plastic pipe, with a recessed lid and latches that could be turned with a coin but not by a bear's claws or teeth.

At the zoo Coffey emptied a can of sardines into the canister, smeared the outside with sardine oil, locked the lid, and threw it into an enclosure with a six-hundred-pound male grizzly bear. The grizzly spent twenty minutes pawing, biting, and standing on the container before giving up. Next, Coffey put the container in an exhibit containing two three-year-old Kodiak brown bears weighing about seven hundred pounds each. The young and energetic bears pawed at the canister, bit it, wedged it into corners, and stood on it. Then, standing on their hind legs, they hurled the canister at the concrete walls and floor. After forty minutes, they lost interest.

Finally, Coffey dropped the canister into an enclosure containing a full-grown 1,400-pound male Kodiak brown bear. After trying other things, the giant bear settled on the strategy of nimbly balancing his huge bulk on his front paws, pushing down on the canister until, after ten minutes, Coffey heard the container crack. The Kodiak then tore it open to lick out his tiny, fishy reward. Since this bear weighed seven or eight times as much as an average Sequoia black bear and over twice as much as many Rocky Mountain grizzlies, Coffey told Werner that in his opinion the container was ready for production.

Within two decades after that day, front-country lockers and backcountry food storage canisters utterly transformed the behavior of bears in national parks and wilderness areas. By 2015, Yellowstone had over 4 million visitors, yet food-conditioned bears were so rare as to be nearly unknown. In the end, all of the effort expended by generations of rangers to move bears around would have been better spent on inventing and deploying food lockers and backcountry food canisters.

THERE WAS SOMETHING in Graber that was the opposite of Cole, and it would prove valuable to him, although he didn't see it that way in the early eighties: David Graber was plagued by doubt. The certainty that Glen Cole seemed to possess once he set out on dump closure and natural regulation was an attribute Graber simply didn't have. Graber worried about making the wrong decision. He wasn't sure he knew enough, or anyone did, to properly manipulate nature.

In the spring of 1983, David Graber telephoned Starker Leopold in

Berkeley to share some doubts about the philosophical basis for Sequoia and Kings Canyon's prescribed burns. After publication of Bonnicksen's article in *Ecology,* Boyd Evison had asked Graber to feel Starker out about writing an updated version of the now twenty-year-old Leopold Report. Graber followed up the telephone conversation with a letter to Starker, enclosing a copy of Bonnicksen's paper. He asked for a meeting between himself, Parsons, and Starker, and he laid out some of the issues that were bothering him.

"If Indians played an important ecological role as burners . . . the Park Service will find itself with the task of simulating Indian burning in perpetuity, yet very possibly without the scientific data necessary for a realistic simulation," wrote Graber. "If Indian burning must be simulated, must also other ecological roles played by Indians, as predators for example, be likewise simulated?" Even if the Park Service were able to duplicate the exact role of Indians as burners in 1850, Graber went on, the Native culture the agency was simulating would be static, frozen in the 1850s. "[A]ren't we imposing the assumption on our simulated Indians that they would forever be trapped on one rung of their own cultural ladder, never to abandon hunting deer and collecting acorns for the greater joys of snowmobiles and Pacman?" he asked.

Starker wrote back that he would be happy to get together, but he warned Graber that the degree of refinement Graber and his fellow Sequoia scientists were seeking was beyond that which he had put into the Leopold Report, and he said that the younger men probably knew more about the subject now than he did. A lunch meeting was arranged at the UC Berkeley Faculty Club, and in the first week of June, Graber, Parsons, Tom Nichols, and Boyd Evison left Ash Mountain to see the oracle.

The Faculty Club, Bernard Maybeck's 1902 tan stucco Arts and Crafts landmark, was nestled in the live oaks of the Berkeley campus. The interior was done in unfinished redwood, with a ceiling supported by timber trusses, the purlin ends of which were carved into wolves' heads. The effect for the younger men was like eating with a favorite uncle at a European hunting lodge. Starker was jovial, but he gently chided the men who had come to see him for abdicating their responsibility to manage. Nature would not stroll back to health on its own. They were in charge, and they needed to make their best judgments based on the best information they had and get on with it.

Graber asked about a new Leopold Report. Leopold laughed. "Starker said there would be no second coming," Graber recalled later. If the Sequoia scientists wanted a new Leopold Report, they would have to write one for themselves, said Starker. He thought Graber and his colleagues were all too afraid of making decisions that would always be somewhat arbitrary. Much of what park managers had to do was guesswork, said Starker, but it needed to be done anyway.

"If an area is ready to burn, it makes little difference to me whether the fire is set by lightning, by an Indian, or by Dave Parsons, as long as the result approximates the goal of perpetuating a natural community," Starker explained in follow-up letter to Boyd Evison after the meeting.

"If too many grazing animals are destroying vegetation, they should be reduced in number, by predators if possible, if not, by trapping or shooting," he continued, in an obvious reference to Yellowstone.

"I am much less afraid of this kind of decision making," Starker went on, "than I am of adopting inflexible rules of conduct—such as, '. . . we must wait for lightning to start the fire; we must wait for a mountain lion to work over those deer.' Our parks are too small in area to relegate to the forces of nature that shaped a continent."

Ten days later, a volunteer from the Sierra Club visited Starker at his office to record an oral history. Fog off the ocean formed a white ceiling over Berkeley. Leopold reflected on the meeting at the Faculty Club.

"[S]ome of the Park Service biologists, including Dave Graber—who was one of *my* products, one of my own kids, and a damn good boy—they're uneasy about arbitrary decisions," Leopold mused. "You decide to cut down a tree; who's to decide which tree to cut? Should you cut any tree at all?

"And they'll all go for the idea of letting natural fires run. If lightning starts a fire, then, this is something that has to do with God, and you didn't have to make a decision. But they were really concerned in a genuine way with arbitrary decisions that have to be made. As soon as you move into management, you're going to have to manage for something, you're going to try to re-create it, try to maintain a given type of ecosystem. And I still defend it. Okay, so you make some arbitrary decisions. So what? They may be arbitrary, but that doesn't necessarily mean they're capricious, as long as your objective is a goal of what you construe to be a natural ecosystem," Leopold concluded.

Two things resonate in Leopold's words. The first is that, in saying "what you construe to be a natural ecosystem," he saw a growing imprecision in the use of the word *natural* to define the goals and techniques of preserving wild places. To *construe* means to interpret, and as scientists struggled to sort out which phenomena in national parks were "natural" and which were the result of previous human actions, Starker meant to say that what constituted a natural ecosystem might be at best an educated guess, subject to revision, and be acknowledged as such.

The second point, and perhaps the most instructive for the young scientists who sought out Starker in the waning days of his life, was that, notwithstanding their incomplete knowledge, they would have to *manage,* to do something, in order not to see their parks degraded. At twenty years old, the Leopold Report may have looked faded in its call to make parks "vignettes of primitive America," but it was not the least bit obsolete in seeing how compromised parks were by their limited sizes and poorly drawn boundaries, by what was going on around them, and by previous tinkering. Starker was saying that to be a guardian, you must be a gardener.

ALTHOUGH STARKER NEVER admitted that he regretted having supported Cole and Anderson, his involvement with the grizzly took on an altogether different tone soon after the September 1972 bear management meeting at which the Craigheads presented the initial findings of their population modeling, which suggested a perilous decline in the Yellowstone population. Upon hearing that the Secretary of the Interior had requested that the National Academy of Sciences look into the bears' status, in February 1973 Starker wrote the National Academy's president to emphasize how important he thought it was that the academy take a serious look at the Craigheads' data on the effects of Cole and Anderson's program.

When Glen Cole left Yellowstone for Voyageurs National Park in 1976, he invited Starker to come fish with him there. Starker never went. By 1977, after the Court of Appeals reversed the Walkers' victory and the case began to fade into obscurity, Starker had grown thoroughly disillusioned with Park Service bear management. In an assessment requested by a new Park Service director, he reported that the chronic failure to warn and educate the public about bear danger—noted by judges in both the Claypool and the Walker cases—had not been remedied. Further, close monitoring of bear sightings and closure of areas where bears and people might come in conflict were still not being practiced to the extent required. Frustrated, in September of 1977 Starker wrote to a colleague that he was trying to "impress upon the Park Service the importance and seriousness of this whole bear issue."

In 1980, in two separate attacks, grizzlies killed three people at Glacier National Park. Early that year, Mary Meagher sent Starker a paper she was about to present to a meeting of bear biologists, on Yellowstone's successful restoration of a natural population of grizzlies. In an accompanying note, Meagher wrote: "In essence, we've done what we set out to do." Starker replied, expressing skepticism about Meagher's claims of success for the program Cole had developed and implemented. The grizzly population had failed to bounce back after the 186 unnatural deaths between 1968 and 1972, and Starker wrote that while he was happy to have Meagher's paper, he would have liked to get from her "data and supportive statements that the population is not in trouble [the underline is Starker's]." By June of 1981, Dr. Richard Knight, the Park Service's own biologist who had been put in charge of the Interagency Grizzly Bear Study Team eight years earlier, stated publicly that the sudden dump closures and the many deaths of bears associated with them had been a mistake.

Since long before recorded history, human civilization had been manipulating nature. Harry Walker had grown up in the ancient tradition of agricultural husbandry, learning to care for a pasture that had been a riparian hardwood forest before his father cleared it. At times he had been elbow deep in a cow's uterus, easing the birth of a calf. And yet, wildness—the unpredictable, the unforeseen—kept breaking through, and it was Harry's elbow that precipitated his departure from husbandry and his arrival at Yellowstone in the midst of a grand experiment in retreating from manipulation.

What Starker Leopold was trying to teach Graber was not to master and control everything but, instead, to remember that because human hands were always unintentionally doing something to nature, they ought to do something carefully planned as well. Starker spent his entire career saving wild land and creatures from the yoke of utility, and if he differed with Cole and the Muries on the extent to which biologists should manipulate the wild in order to repair it, he was united with them in opposing total domestication of the earth. Indeed, an appreciation for the untamed unified everyone in this story, even Harry Walker and his mother during the troubles before he left home.

One evening a few months before Harry's death, he was driving Louise back from shopping in town when he asked if they could detour to a place that was special to him. Louise agreed, and they passed the farm and continued down Old Downing Mill Road, turning off on the dirt track to Cobb's Bottoms, a lowland hardwood forest that had not been cleared for pasture, where Harry liked to hunt and fish. At the road's end Harry turned off the car, and the two of them sat, wordlessly, with the windows open, feeling the affection of a mother and a son, watching the stars between the trees through the windshield, and listening to the deep chorus of bullfrogs in the warm Alabama night.

EPILOGUE

THE ADVICE STARKER gave Graber and his fellow scientists in June of 1983 was his last. It is not known whether he had a premonition, but in the month after recording his oral history, he had a heart attack, which he survived. Then in August of that year he had another, which killed him.

By the time of Starker's death, in the Rocky Mountains, a fifty-year-old philosophy professor with multiple Ivy League degrees, Alston Chase, had sequestered himself on a Montana ranch to write a withering critique of natural regulation. Published in 1986 as *Playing God in Yellowstone: The Destruction of America's First National Park,* the book became a *New York Times* bestseller. Its villains were Yellowstone biologists Glen Cole, Doug Houston, and Mary Meagher.

Chase got much of his analysis from Charles Kay, a Utah State graduate student who was writing a dissertation refuting natural regulation under Frederic Wagner, a professor of wildlife management and fierce critic of the policy himself. In Kay's analysis, Yellowstone had been sold to the public as a cornucopia of wild animals. In fact, Kay said, some of the first white explorers had not found a lot of game there. Some nearly starved for lack of animals to shoot, he claimed. Elk might not even have been native to Yellowstone, asserted Kay, citing a paucity of elk bones in archeological digs. After the wildlife "holocaust" of the 1870s and '80s muddied the waters about what exactly had been there, wildlife ranching and the winter feeding of elk, bison, deer, and antelope, along with the destruction of wolves and cougars, remanufactured Yellowstone into a romantic image of a frontier Eden with endless carnal sustenance for settlers. Swollen elk herds pushed white-tailed deer and antelope into decline and ate through the willows along stream banks, causing beaver in the Northern Range to all but disappear.

Then, with the park in ecological uproar, in 1967 Cole and his people decided to abdicate their responsibility to repair the place, leaving the ecology of Yellowstone to its own devices under natural regulation. Chase characterized natural regulation not as a product of science but as part of a pantheist religious revival promoted by fuzzy-headed Earth Day shamanists whose sacrament was the contemplation of the mysterious power of nature.

Chase detailed the Park Service's killing of grizzlies, the agency's dismissal of the Craigheads' carrion feeding during dump closure, and what he described as the starvation of grizzlies for lack of enough natural foods in the decade that followed. Kay told Chase he suspected that one factor adding to the grizzly's decline after dump closure was the elks' browsing of bushes that bore late-summer chokecherries and berries, key foods to pack on weight for hibernation during the bears' autumn feeding frenzy.

What natural regulation's failure ought to teach us, argued Chase, was that we could not retire from active involvement with nature. We ought to accept our role as stewards, not pretend we can stand back from the remaining shards of much larger ecosystems, clinging to a mystical belief that some great, intact force of homeostasis—the "balance of nature"—would repair everything. Chase was right about some of this, but *Playing God in Yellowstone* was also a bomb lobbed leftward in the culture wars. Chase characterized Yellowstone biologists and eastern-emigrant environmentalists—"dressed in their neatly creased designer jeans, Orvis flannel shirts, and L.L.Bean Maine rubber moccasins . . . a kind of unisex uniform of the trust-fund cowboys of the Rocky Mountain West"—as secret members of a cult of pantheist religion. Chase himself, however, was not without a cultural agenda. He was a conservative ideologue who later wrote a book about how the Marxist intellectual establishment in American universities had created the arch-terrorist

Ted Kaczynski, known as the Unabomber. *Playing God* became part of the so-called Sagebrush Rebellion against the Park Service, government in general, and environmental regulation of all kinds. Chase was widely quoted in documents such as a white paper on how the government was destroying the environment, written by the director of the Maguire Oil and Gas Institute at Southern Methodist University. Released in an updated paperback edition in 1987, *Playing God* set the ideological table for the media's uncomprehending treatment of landscape-level changes in Yellowstone in 1988.

IN THE FIFTEEN years after its first lightning-strike fire was allowed to burn under Jack Anderson and Glen Cole, Yellowstone had a harmonious honeymoon with natural fire in which less than 2 percent of the park burned over. With cooperative natural-fire programs in place for neighboring national forests, this tended to confirm the impression that Yellowstone was big enough to let fire do its work unimpeded.

However, by June 1988, Yellowstone was having its hottest, driest year on record. Ten days before NASA's James Hansen delivered his epoch-making testimony that climate change was under way before the Senate Energy and Natural Resources Committee, lightning struck in a national forest wilderness near Yellowstone's northeast corner. Other strikes kindled blazes the day before and the day after Hansen's testimony, and new ignitions continued into July.

In consultation with neighboring forest supervisors, Yellowstone's superintendent, Bob Barbee, at first stayed the course on natural fire. But by early July, he and his fellow land managers were seeing fire behavior that was well outside their comfort zone, so on July 15, with more than eight thousand acres involved, they made the decision to suppress all new ignitions and began calling in resources to put down the blazes they already had. Truck- and planeloads of men, women, and equipment rolled in. By the end of the July the moisture content of fine fuels such as twigs and pine needles had dropped as low as 2 or 3 percent.

Coarse fuels such as fallen trees and limbs were around 7 percent, about the same as kiln-dried lumber. The danger zone in large fuels was over twice that.

The principal way wildland fire is fought is by clearing strips of land of all fuel, thereby interrupting the progress of the fire. However, once you reach the point where fire-created winds are blowing superheated air, sparks, and flaming pieces of wood for miles, all bets are off. That summer at Yellowstone, fire jumped highways, crossed the Yellowstone River, and leapt half a mile ahead of itself on the wind, trumping all normal methods of firefighting. The fires "kicked our ass from one end of the park to the other," one firefighter told the historian Hal Rothman. In the middle of the morning on August 20, the wind around Madison Junction picked up to forty miles per hour, with gusts up to seventy. Trees snapped like matchsticks, the fire roared like a freight train, and flame heights reached two hundred feet. An observer in a spotter plane saw mushroom clouds like those of nuclear explosions rising from Yellowstone.

The fires closed in on Old Faithful the week after Labor Day. Most of the buildings there were made of wood, and the Old Faithful Inn's giant roof was covered in wood shingles—kindling, essentially. The fire advanced from the bluffs on the west side of the valley. Jack Ward Thomas, who later became chief of the Forest Service, had been sent to assess Yellowstone's fire policy for the National Academy of Sciences. He found himself lying in an Old Faithful parking lot, clutching his hard hat as flaming pieces of trees flew overhead. A spot fire started on the roof of the Old Faithful Inn. Employees heroically put it out. Sprinklers had been set up on the inn's roof and in the surrounding meadows. Fire burned on all sides of the village. North of Observation Point, the location of Harry and Phillip's campsite burned over, but the village was saved. By the time the fall rain and snow had put them out, the 1988 fires had burned a million acres. The largest recorded fire in Yellowstone's history up to that time totaled eighteen thousand.

BY THE TIME of the fires, the northern elk herd numbered twelve to fifteen thousand. The fires had burned some forage, and the following winter was a bitter one; many died. But the fires stimulated new growth in the spring, supporting lots of calves, and the population remained high. Those who believed that aspens had been held back by fire suppression rather than by elk were pleased to see aspen seedlings all over the Northern Range following the fires; few of them got more than knee-high before they were eaten, however. What willows existed were stunted, and beaver were still largely missing in the park's north.

From outside the park, the chorus against natural regulation intensified amid the critical rhetoric about the fires. Congress appropriated money for the Park Service to make a study of the Northern Range. Not surprisingly, the results of the in-house study supported the present policy of natural regulation. Appraisals from outside the Park Service were less enthusiastic. An ad-hoc committee within the Wildlife Society published their findings in 1995, which were highly critical of natural regulation. The Senate Subcommittee on National Parks held hearings on Yellowstone's management of its grazing animals in 1996. A study by the National Research Council of the National Academy of Sciences, published in 2002, found the word *natural* in *natural regulation* severely wanting as a goal for ecosystem management. "In view of the profound changes that have occurred," said the NAS committee, ". . . it is no longer possible to have an ecosystem that is truly natural . . . containing the same numbers and distributions of all the species of plants and animals that were there before European settlement."

ROCKY MOUNTAIN WOLVES were all but extinct—a few straggling in from Canada in the north—when the Endangered Species Act was passed in 1973, and they were listed under the act in 1974. In a broader ruling in 1978, the Fish and Wildlife Service listed wolves as endangered in the forty-eight contiguous states, except in Minnesota. The agency filed a recovery plan in 1980, which called for transplanting wolves from

Canada back into the United States. The plan's authors considered two places large enough for the wolves to have a chance: the 2-million-acre complex of Forest Service wilderness areas in central Idaho, and the Greater Yellowstone Ecosystem. Fifteen years of political wrangling followed. The return of wolves was unpopular in rural areas of the West and among Republicans and popular in urban areas and among Democrats. Wolves were most unpopular in the states where the recovery plan called for them to be released, and most popular in states where it didn't.

In November and December of 1994, thirty-one wolves were captured in Alberta and flown to the United States, where they were held, awaiting a judge's decision on an injunction to block their release. When the decision came in January, wolves were moved to three locations on the Northern Range. There they were kept in pens, fed on wild meat, and allowed to acclimate to the sights and sounds of Yellowstone for three months. Then they were let out. The census of Northern Range elk that winter was 16,791 animals, which meant that after calving that spring, the herd would top 20,000. The park was a meat locker. The wolves thrived. Sixteen more were introduced in 1996. The park's biologists named one of the first groups of freed wolves the Leopold pack, after Aldo Leopold.

As wolves multiplied, the population of Northern Range elk plummeted, from some 19,000 in 1994 to only 4,844 in February of 2015. In a cooperative effort between the park, the State of Montana, and the Rocky Mountain Elk Foundation, more winter range had been acquired north of the park, and even before the arrival of wolves, an increasing number of elk had been migrating north. In 2015, only 1,134 northern elk spent the winter inside Yellowstone. By 2001, willows, aspens, and narrowleaf cottonwoods were sprouting up in the Northern Range. With the willows came species of songbirds that frequent riverside brush and had been scarce, if not absent, before the recovery.

Prior to wolf reintroduction, the pronghorn population had been declining. Coyotes are the chief predators of pronghorn fawns. When wolves were put back, the coyote population took a nosedive and pronghorn rebounded. Winter wolf kills also benefited grizzly bears, which fed on carrion when they emerged from hibernation in the spring. As willows reestablished themselves, the Northern Range's missing beaver began to return. When the Yellowstone wildlife biologist Doug Smith did his first survey, in 1996, he found only one active beaver colony. By 2005 there were ten, by 2015, twelve. However, wolves may not be totally responsible for the changes predation brought to the Norther Range. Mountain lions were hunted out of Yellowstone during the predator-control era, but they are coming back now.

Today, visitors see far fewer elk in the Lamar Valley. What you see now are lots of bison. Before natural regulation, the Park Service limited bison herds to between 800 and 1,350 animals, removing some to create herds elsewhere, shipping others to slaughter, and shooting those found to be infected with brucellosis. Called undulant fever in human beings and Bang's disease in animals, brucellosis arrived in the United States from Eurasia and probably passed from domestic milk cows kept at Yellowstone in the early days to the park's bison. It was the fear of brucellosis—which can pass from cows to people in dairy products—that left Wallace Walker's milk-delivery truck gathering dust on his farm after Alabama outlawed the sale of raw milk. From the 1930s on, brucellosis was systematically eradicated in the United States by the killing of entire herds of infected animals. Following the success of this program, the only disease reservoirs left were Yellowstone bison and elk.

From 1968 onward, as with elk, the bison population was left to find its own level. For a long time, it seems that competition for forage with vast numbers of elk limited the expansion of bison. When elk declined after the reintroduction of wolves, bison really took off. Mary Meagher, the park's longtime bison biologist, observed them colonizing new areas of the park during the winter, traveling on packed trails laid for the park's controversial snowmobile winter recreation program. She believes the trails had a role in the story.

When bison had the whole Great Plains, they were nomadic animals. However, under agreements that Yellowstone National Park was forced to make with the Department of Agriculture and with neighboring states over fears about brucellosis, bison were required to remain in the park. As the herds expanded, in the winter, larger numbers of them began trying to migrate across the park's northern boundary into the Yellowstone River Valley. Between 1985 and 2014, 7,800 bison were killed by agents for the State of Montana and by hunters authorized to take them under a state program.

By 2015 the Yellowstone bison were estimated to number 4,900. "The park has been turned into a bison factory," said Mary Meagher. In her eighties the retired biologist, long a staunch supporter of natural regulation, thought the only thing that could have stopped the population explosion was a major shooting inside Yellowstone before it got out of hand. But the public has even less stomach for that now than it did for elk reductions in 1962, and the only politically feasible control at present is to use the agreement with Montana and the Department of Agriculture to justify killing buffalo when they walk across the Montana State line. In 2015 the Park Service announced a plan to kill a thousand that way, employing a legal hunt in Montana and capture and shipment to Indian tribes for slaughter.

NOTWITHSTANDING WHAT MEAGHER called a "change in the grazing system" from one based on elk to on one based on bison, intensive study by Yellowstone historians Paul Schullery and Lee Whittlesey suggest rather conclusively that *Playing God in Yellowstone* author Alston Chase and his informant, Charles Kay, were wrong about a scarcity of large animals in the early days, before—according to them—the Park Service turned Yellowstone into a wild animal show and destroyed the Northern Range.

In 1991, Schullery and Whittlesey set out to see what the historical record really said. Over the next twenty-five years they accumulated every eyewitness account they could find of the presence and abundance of wild animals in and around Yellowstone, from the end of the 1700s to the 1880s, a collection that eventually encompassed over five hundred sources. The results demonstrate that Yellowstone was not, as Chase and Kay concluded, the sort of place early explorers were in danger of starvation for lack of game. More typical of the actual conditions was a day in August of 1837 when trapper Osborne Russel crossed a pass and descended toward Yellowstone Lake, where, he wrote "we found the whole country swarming with elk." Many other sources demonstrated that this was not just fur trappers' hyperbole.

Notwithstanding the long dispute over how many elk ought to be on the Northern Range, one of the great lessons of the reintroduction of wolves was that elks' effect on vegetation could not be measured by their numbers alone. In the late 1990s two scientists from Oregon State University, geographer William Ripple and hydrologist Robert Beschta were doing research on the cascading effects of wolf reintroduction on other components of the system of life on the Northern Range. In the summer of 2001, Ripple was working in the Lamar Valley when he noticed that willows and narrowleaf cottonwoods were growing back profusely in some types of terrain and not in others. Ripple, Beschta, and their graduate students analyzed the difference by terrain type, then compared the locations where vegetation was doing well to other researchers' data about where elk had been killed by wolves. What they found out was that predation was happening in the same places where willows and aspens were growing back: holes, gullies, and tight spots against moraines, where elk couldn't see wolves coming and their escape routes were limited.

Ripple explained the phenomenon with a hypothesis he called "The Ecology of Fear," suggesting that elks' effect on vegetation was entirely different when they could no longer stroll with impunity into thickets or gullies and put their heads down to eat. He pointed to 1968, when, after the final year of herd reductions, the northern herd was down to around four thousand animals. Yet willows, cottonwood, and aspens hadn't recovered. Then again, in the early 2000s, with over twice as many elk in the presence of wolves, vegetation was coming back in places it hadn't in the late 1960s. Ripple, Beschta, and other researchers studying these spreading effects from predation, called "trophic cascades," suggested that aspen forests and even the courses of rivers—because bankside willows and beaver dams affect stream velocity and bank erosion—could be formed by predators. Natural regulation had been turned on its head. This research suggested that plants didn't control animals from the bottom; predators controlled plants from the top; or at least the actual process was a combination of the two.

TODAY, AREAS OF Glacier and Yellowstone National Parks are sometimes closed when grizzlies are present, as the Craigheads recommended. When a grizzly is seen on a trail, that information is posted at the backcountry permit offices and dispensed to the public. In areas where bears are frequently seen, such as the northeast portion of Glacier National Park, the Park Service employs roving monitors trained to keep bears and people apart. Bear management technicians now carry shotguns loaded with rubber riot rounds. When young bears begin to be interested in people or campgrounds, and before they get their first food reward, they are "hazed" to increase their wariness of human beings and developed areas. Incidents involving black bears where food conditioning was the problem are now rare. With grizzlies, fatal attacks, although very rare, continue to occur. Including Julie Helgeson and Michele Koons, Glacier National Park has had eight, in which a total of ten people have died. Yellowstone has had nine fatalities, including Harry Walker. Five additional deaths have occurred just outside the park. In addition to those related to food conditioning, there are three other categories of attacks: Grizzlies sometimes attack when a victim inadvertently approaches a carcass on which the bear is feeding. Older bears in poor nutritional condition, like the ones that killed Harry Walker and Michele Koons, have been known to prey on human beings. And females will attack to protect their cubs, as was the case with the bear that mauled Smitty Parratt.

The latter was probably the case in October of 1992, when a man backpacking alone was killed by a grizzly within a couple hundred yards of where Julie Helgeson and Roy Ducat were attacked, below the Granite Park Chalet. The park's bear management biologist flew in by helicopter, found the man's camera and tripod, and followed a trail of blood and personal items to the body, which had been partially consumed. Leaving the area, the biologist was charged by a sow with cubs. He spent the night in the trail-crew cabin where Roy Ducat had sought help, and when he returned the next day, the body had been moved and more had been eaten. The sow and cubs were destroyed.

Judge Hauk's suggestion that when bears are captured for hanging around people they be equipped with a radio collar so biologists can keep an eye on them has become standard operating procedure in some areas. Yosemite has fifteen collars equipped with both radio- and satellite-tracking devices, like the one Frank Craighead dreamed of. The devices upload three location fixes an hour. The park also employs bear-locating devices, which are fixed sensors that can be deployed in an attractive place such as a campground and adjusted to report the approach of a bear at a given range. The locators can even send a text message to a biologist's cell phone. At Yosemite, a biologist can check on her tablet in her living room the locations of all collared bears before she goes to bed.

EVEN AFTER THE grizzly was listed as a threatened species, in 1975 things did not immediately turn around. Grizzlies have a very low reproductive rate; females of breeding age produce cubs only every two or three years. In the seventies and eighties there was a lot of jostling between government agencies, timber interests, ranchers, and developers out of fear that the designation of critical habitat would stop development, agriculture, mining, oil and gas extraction, and forestry. By 1990 the estimated Greater Yellowstone grizzly population had dipped to ninety-nine bears, and then it began to creep up again. The Greater Yellowstone Ecosystem, now calculated at as much as 18 million acres including two national parks, three wildlife refuges, and portions of six national forests, as well as private land, supports an estimated 655 to over 1,000 grizzly bears.

Other than a brief hiatus during the mid-eighties, Sequoia and Kings Canyon National Parks' prescribed burning has continued since 1969, and some of the young sequoias that sprouted after the first burns are well on their way to joining the ranks of the giants. From the time David Graber came to Sequoia in 1979 to his retirement in 2014 as the chief science advisor for the Park Service's Pacific West Region, his office remained in the old superintendent's residence at Ash Mountain. His career was characterized by the practice of Starker Leopold's assertive manipulation of nature. He consulted for years on a major project to restore Channel Islands National Park. Called "America's Galápagos," the Channel Islands are the home of plants and animals seen nowhere else in the world. A number of these species exhibit giantism and dwarfism, which are characteristic of genetic isolation on islands. Found there are a pygmy fox and an electric-blue jay about one and a half times the size of the scrub jays on the mainland.

But there were also a lot of feral farm animals. Sheep, goats, and pigs had been introduced to some islands, and deer and elk to others, for sport hunting. The Nature Conservancy, which had purchased land on the largest of the islands, and the Park Service set out on a massive program of what animal-rights advocates saw as ethnic cleansing. They shot some thirty-eight thousand feral sheep and well more than a thousand feral pigs, as well as wild goats and deer, all introduced from the mainland.

At another of Graber's consultancies, Point Reyes National Seashore, previous landowners had introduced axis deer from India and fallow deer from Europe. The fallow deer had antlers with spade-like protrusions, similar to those of caribou, with which they plowed up the ground when they were grazing. Rangers began shooting the deer in the 1980s, and later, in cooperation with Graber, they brought in professional hunters. An outpouring of grief and anger from local residents followed. After a time the park stopped the killing and the deer were humanely eradicated via contraceptive implants and trapping.

Other manipulations that Graber favored involved putting things back. His interagency work on the reintroduction of bighorn sheep to their historic range in the Sierra Nevada was a major success. And for most of his career, in line with his mentor, he troubled little about hands-on biology. When a high-altitude frog was desperately endangered by a fungal disease at Sequoia and Kings Canyon, he approved studies by academic researchers and recovery work by Park Service crews that involved capturing hundreds of the amphibians, dipping them in chemical and bacterial solutions to find something to control the fungus, using gill nets and electrofishing to remove non-native fish from their habitat, and relocating the frogs to different areas to build resilience in the populations when necessary.

In 2010, Graber was a contributor to a book called *Beyond Naturalness: Rethinking Park and Wilderness Stewardship in an Era of Global Change.* The word *natural,* on which had been based not only natural regulation but the Wilderness Act and the Park Service's management policies for wild areas, had completely lost its meaning, claimed Graber and his coauthors, a group of scientists associated with the Aldo Leopold Wilderness Research Institute in Missoula, Montana.

Natural, the authors pointed out, had come to mean two divergent things. On the one hand, the word implied that a place was unmanipulated by human beings, its processes guided only by nature. On the other hand, it meant the land still had species and features it had had before the wave of human-caused change that was sweeping the earth—that is, it wasn't covered with cheatgrass, kudzu, Brazilian pepper, Russian olive, Eurasian wild boar, axis or fallow deer, or European brown trout.

These two definitions were now in conflict. In the second case—those landscapes that had the approximate makeup of, say, Yellowstone in 1850—it would take more and more intervention to preserve that state. Yet that would make such places more unnatural if, in the sense of the Wilderness Act, *natural* meant uncontrolled by human beings.

STEPHEN ZETTERBERG NEVER got a penny in legal fees for his work for the Walkers, but he remained regularly in touch with them until shortly before his death, in 2009, at the age of ninety-two. William Spivak served at the US Attorney's Office's Civil Division in Los Angeles for forty-three years. After the Parratt, Vaughan, and Walker cases, he took every opportunity to work on matters involving national parks and national forests. Someone in the Park Service gave him an honorary ranger hat, which he proudly displayed for visitors. He died of cancer at seventy-five in 2010.

Harry Walker's younger sister, Jenny, married a warm-hearted truck driver with a wicked sense of humor named Eddie Whitman. On the night of their wedding rehearsal dinner, one of Wallace's cows dropped a calf, and Jenny and Eddie had to go out and get it so Wallace could get dressed. The young couple took out a mortgage to buy the Walker farm so Wallace and Louise could go on living there until they died, and in 1996 Jenny and Eddie moved into a mobile home next to the old farmhouse so Jenny could care for her parents. Wallace treated Eddie like a son, but Eddie hadn't grown up on a farm. He felt unskilled, and he worried he could never replace Harry. One day when Eddie went to work, he left a note on Wallace's kitchen counter:

> *Wallace,*
>
> *Leave me a list of things that need doing and I'll fix them or make them worse.*
>
> *Eddie*

Wallace's middle daughter, Carolyn; her husband, Walter; and her children moved home from Kentucky, settling on land they bought near the Walker place in Cobb's Bottom, where Harry liked to hunt and fish. Wallace milked his last cow in 1990. When he died at the age of ninety, in September of 2004, the Walkers' family doctor warned them that Louise wouldn't last long without him. She passed away the following month. After her parents' deaths, Jenny tore down the milking barn, chicken house, and sheds, and burned Wallace's hay barn like a funeral pyre. She and Eddie tore down the old farmhouse, replacing it with a lovely, efficient modern house. The spot where the hay barn stood is still visible as a red-clay depression where Wallace's cows used to walk in to eat at the hay cribs. Eddie uses the site as a par-three golf course. One day he broke one of Betty's daughter's windows while teeing off, and he is trying to be more careful. The corn crib that Wallace and Harry built still stands. Wallace's pasture became a wildlife refuge under a conservation easement. The old pine tree fell down.

Betty's daughter, Renee Simmons Raney, grew up on the farm listening to her grandmother's bitter regrets that Harry had died because people didn't understand how to live with bears. For Renee, Harry's death was the family's central legend. She got a bachelor's degree in biology and a master's in environmental biology and found her calling in environmental education. Today she is known throughout the Southeast as a developer and administrator of programs that teach Alabama children to appreciate and live safely with nature, and she has been the recipient of dozens of honors and awards. As this book was completed, she was preparing to teach a workshop on the thought of Aldo Leopold.

During the grizzly controversy, Martha Shell came to be well known to the National Park Service's Washington staff, to Under Secretary of the Interior Reed, and to assorted senators and members of the House of Representatives and their staffs. After the Walker trial, she seems never to have participated in politics again. She kept in touch with the Craigheads and the Walkers, and she tried to write a book on the Walker case but didn't finish it. Her husband, Paul, died in 1991, leaving thousands of color slides of animals, and Martha followed him in 1994.

Jim Brady rose through the ranks of the Park Service to become chief of ranger activities—the so-called chief ranger of the Park Service—in Washington, DC. He retired to Durango, Colorado, one of many places to which Yellowstone elk had been shipped to rebuild herds wiped out during the nineteenth century. Each autumn, Brady set out to fill his freezer with one of those Yellowstone elks' descendants. In the autumn of 2013, the Yellowstone elk almost put Brady on ice when a hunting accident left him hospitalized for three months with broken bones and head injuries. If his wife, Gwen, will ever let him, he plans to shoot another one.

Maurice Hornocker is a professor emeritus at the University of Idaho. The Himalayan tahr, the subject of Graeme Caughley's 1970 paper about how animals come into balance with their environment, became a notorious pest in New Zealand. Adolph Murie's last professional act was a 1974 paper intensely critical of the advent of prescribed burning at Grand Teton National Park, because the fires would be started by humans, not by nature. He died in 1975.

Smitty Parratt maintained a straight-A average in high school and, with one eye and only one complete lung, he lettered as a cross-country runner. He grew up to become a Park Service ranger naturalist, like his father, and retired as the chief naturalist at Wrangell–St. Elias National Park and Preserve in Alaska. On his retirement he became an instructor at a community college in Oregon, where he teaches Spanish, biology, kayaking, backpacking, wilderness survival, and Tai Chi. On summer vacations he hiked the Continental Divide through the Rocky Mountains. In the summer of 2014, he completed the hike by backpacking through Glacier National Park, where he had been attacked as a boy, and where grizzlies still roam.

Galen Rowell became one of America's most famous outdoor photographers and the author of several books. He climbed and adventured all over the world and operated successful photography galleries in Emeryville, near Berkeley, and Bishop, California. On August 11, 2002, Galen and his wife, Barbara, were killed in the crash of a private aircraft north of Bishop while returning from teaching a photography workshop in Alaska.

After Harry Walker's death, Vikki Schlicht was married and divorced four times. She left Montana and settled in Alabama not far from the Walkers, where she works as a midlevel manager for a large corporation.

Family members say that Phillip Bradberry never got over what happened to Harry Walker. He drank and drifted, and when he became too weak to continue he was taken in by a daughter he had abandoned earlier in his life. He died of cancer at her home in January of 2014.

Frank Craighead continued to live on the Jackson Hole compound where he and his brother had built their log cabins after World War II. He wrote popular articles and two books and fly-fished with his sons, daughter, and friends. Yvon Chouinard, the Yosemite climber and founder of Patagonia, became friends with Frank after he and his wife purchased a home in Jackson Hole. Like Frank, Chouinard was an avid fly-fisher. Chouinard's climbing career had begun when as a teenager he happened upon Frank and John Craighead's 1937 article on falconry in an old copy of the *National Geographic.* As the Craigheads had done, he set out to teach himself the sport and began climbing cliffs in Southern California to get raptor chicks. He quickly became enamored of the climbing itself and gave up falconry. In the early 1980s, Frank Craighead was diagnosed with Parkinson's disease, and over the next two decades he slowly lost the ability to walk, speak, and take care of himself. Shortly before his death, Yvon Chouinard took him fishing for the last time. He carried Frank on his back into the National Elk Refuge, where the southern Yellowstone elk still winter. They fished Flat Creek for whitefish on nymphs. Then he carried Frank back. Frank died in a nursing home in Jackson Hole in October of 2001.

The first time I saw John Craighead, he was ninety-eight years old. It was December of 2014, in Missoula. The sky was a sheet of white, and the ground was a sheet of white. Craighead was wearing hiking boots, jeans, and a black down jacket with pieces of pink surveyor's flagging as zipper pulls. He had a hiking pole in each hand and was walking down an incline of old snow that had turned to ice, toward his home on the slopes of Mount Dean Stone.

I had first approached John Craighead's children about coming to see him a year and a half earlier, and no one was very keen about it. Eventually they consented. Craighead had dementia. But when I saw him walking on that ice, there was something highly focused and intelligent about the way he moved. Not all the body's knowledge is in the brain. He placed his feet with expert deliberation.

I spent five days with John; his wife, Margaret, age ninety-four; their son, Johnny; and, sometimes, their daughter, Karen. In his later years, John had taken up painting, and in the subjects of his paintings he returned to his early fascination with birds. The paintings hung in the kitchen. They were full of bright colors. As John's mental status had changed, they had become more stylized, almost primitivist, but complex in their ornamentation. They are exuberant and quite beautiful.

For lunch, Johnny fixed us soup and sandwiches; however, his father would never sit all the way through eating. He was restless. Sometimes he sat in a chair in the living room as Margaret dozed nearby on the couch. But in general he could not be kept sitting and could not be prevailed upon to stay indoors. Johnny said his father had a tendency to go outside in the most miserable conditions and stay out there.

So his children had put up a tepee in the front yard. With Karen's help, John had decorated the tepee with bright, primitivist paintings of birds and animals. Inside, in the center of the tepee, Johnny had built a stone fire ring, next to which he piled stacks of kindling. In the afternoon, when it was about ten above zero, John would go out there, Johnny would help build a fire, and John would sit in a traditional Native American folding chair staring into the fire and feeding it sticks.

One day I asked Johnny when John's decline had started.

"With the bear controversy," he answered, without stopping to think about it. He was a stocky man in his sixties, short, like his father.

I explained that I was referring to the present dementia.

"No, that's what I meant," said Johnny. "It started with the bear controversy. He was taking tranquilizers, and then the doctor put him on something else—I don't remember what it was called. But we were never the same after that."

AFTERWORD

THE POWERFUL RESENTMENTS that carried into the courtroom in *Martin v. United States* have receded into history with the deaths of Frank Craighead, Glen Cole, Jack Anderson, and Starker Leopold. Those who are still alive as of this writing, Mary Meagher and Doug Houston, went on to make important contributions to science.

In the end, all the scrutiny of Yellowstone, the pressure on Cole, Meagher, and Houston, and the suffering of the Craigheads bore good fruit. Today there is a two-story building full of scientists in Mammoth Hot Springs, the Yellowstone Center for Resources, and a beautiful, glossy journal in which to publish their findings, *Yellowstone Science,* paid for by a nonprofit association and written for the general reader.

But the questions behind those struggles from the Big Kill of 1962 that triggered the Leopold Report—How much should we respect nature's autonomy? How much should we try to manipulate and control it to save it? Do we know enough to risk doing it? And what happens if we get it wrong?—have not been conclusively answered, nor should they be. They are more useful as questions that ought to be asked every time we face any decision about preserving life on earth than any answer we can give today.

What Starker Leopold told David Graber and his colleagues at their last meeting before his death in 1983 is still true. Only fools are comfortable operating with less than complete knowledge in a contingent world, but we have to get used to it. Things will have to be done, and we must learn from our mistakes as we go. There should always be a certain reticence, a deep and worrisome doubt. It is when we do away with it that we are most in danger. Glen Cole did what he believed in, and, although he was wrong, his respect for the autonomy of nature was commendable. If there was anything wrong with him, it was his certainty that things would be okay in the middle of an unsustainable loss of grizzlies. The grizzlies were okay in the end, but more because of the uncertainty of others than because of his certainty.

In 2012, the Park Service did what the Sequoia and Kings Canyon superintendent Boyd Evison had wanted Starker Leopold to do almost thirty years before: it published a new Leopold Report. Titled "Revisiting Leopold: Resource Stewardship in the National Parks," the report, not surprisingly given present challenges, reaffirmed Leopold's suggestion that management of national parks "may involve active manipulation of the plant and animal communities." In the face of climate change's already visible effects on national parks and wilderness areas, the Park Service has started a few small-scale, tightly-controlled experiments with assisted migration, the practice of helping plants and animals move to new locations where they might better survive the dislocations to come. But in general, even today, there is no great appetite within the service for taking liberties with nature. Outside the ranks of Park Service scientists, however, climate change has loosed grand visions of gardening the whole world, and some authors have suggested that such restoration efforts as the reintroduction of wolves or removal of exotic species are now pointlessly sentimental. Since we have become the biggest act in town and are changing everything anyway, goes this line of thinking, why not admit it and take command? However, given that our grandest planetary effects—our hijacking of the nitrogen and carbon cycles and the wave of human-caused extinctions—are for the most part entirely accidental, to say that the catastrophe we have caused has prepared us to take command of life on earth is akin to saying that being involved in a traffic collision qualifies you to be a highway safety engineer.

Looking back on the struggle between doing and not doing, many of the manipulations that have proved most satisfying in the long run have involved putting things back: a bit of prairie at the University of Wisconsin-Madison Arboretum, fire in the sequoias, wolves to Yellowstone, and bighorn sheep in the Sierra. But even in a clear case of putting things back, an ethical dilemma has emerged.

Isle Royale National Park, an island in Lake Superior, was not pristine when the Park Service took it over. It used to have woodland caribou and lynx. It doesn't anymore. It had been mined, burned to find places to mine, and inhabited by humans. In years when the lake froze over, which used to happen more, moose had wandered out there. In the 1940s, wolves followed, and a tiny, somewhat simplified ecosystem was created that became the most famous and longest-running research project in predator-prey relations in the world. It was there that the great twentieth-century American predator biologists such as David Mech, Rolf Peterson, and Doug Smith, who managed wolf reintroduction at Yellowstone in 1995, learned their trade. But canine parvovirus, an exotic pet disease, snuck onto the island, and wolves have suffered. By the spring of 2015, there were only three left. Without wolves to keep their numbers in check, Isle Royale moose can be a gruesome spectacle, gaunt with malnutrition and eaten alive by thousands of ticks. Yet some biologists feel that bringing wolves back would put an end to nature's experiment on Isle Royale: both moose and wolves showed up on their own, and what has happened since is in some sense "natural." Further, in addition to its status as a national park, Isle Royale has been designated wilderness, and there has been resistance—along the lines of Howard Zahniser's imprecation that in wilderness we should be guardians not gardeners—from the wilderness community to wolf transplants.

Everglades National Park, with the complex of other National Park Service areas around it, including Big Cypress National Preserve, has become a case study at one extreme end of the intervention spectrum. Everglades is the largest remaining tropical wetland in the United States, and everything grows well there. Hundreds of nonnative warm-weather ornamental plants and exotic pets make landfall in North America through the Port of Miami, and among the most insidious that have escaped or been planted since the late nineteenth century are some serious threats to native ecosystems. I spent time in Everglades and Big Cypress on background research for this book and accompanied a pair of hardworking biologists in forays to hunt down giant African monitor lizards and South American tegu lizards introduced by the pet trade. Thousands of acres of introduced exotic plants such as Brazilian pepper, Australian pine, Melaleuca, and European climbing fern have been sprayed with herbicide and attacked with chainsaws there.

Glacier National Park had undertaken a major tree-planting effort to replace about half the Park's whitebark and limber pines lost to exotic white pine blister rust; to native pine beetles, which are moving uphill with climate change; and to climate stress. Whitebarks are not just beautiful to look at—growing up high, they are shaped by the weather into graceful forms like Japanese bonsai—but are also a key element in a web of life that includes the Clark's nutcracker, which feeds on the nuts and flies them around to new homes, and the grizzly, for whom whitebark pine nuts are an important source of protein and fat during hyperphasia, before hibernation.

Some whitebark and limber pines demonstrate a natural resistance to the pine blister rust. The Park Service has been collecting seed from the resistant trees and sending it to a Forest Service nursery in Coeur d'Alene, Idaho, for propagation. Since the program began in 2001, over twenty thousand whitebarks and nine thousand limber pines have been planted in the park. This is hardly natural regulation.

In the East, a genetically engineered chestnut tree that is resistant to the Asian chestnut blight has been developed to fill the missing chestnuts' important niche in eastern forests. Sure, it isn't natural, but what's worse: a new engineered chestnut or no chestnuts at all?

Roger Kaye, a longtime bush pilot and wilderness specialist for the US Fish and Wildlife Service at Alaska's Arctic National Wildlife Refuge, who in his winters earned a doctorate in history at the University of Alaska, has seen what is going on in Everglades and is writing a book contrasting it with the Arctic Refuge, which at present doesn't have an alien plant or animal problem or much need for intervention. His analysis of the intervention-nonintervention spectrum makes sense to me. Kaye says that neither position works everywhere. In places such as the refuge, at least for now, a condition of nonintervention and respect for nature's autonomy can be maintained, along with a healthy reticence to jump in and do things. In others, such as the Everglades, full-on intervention in support of native and endangered species is appropriate.

No natural law requires us to embrace one or the other.

THREE TIMES WHEN I was writing this book, I walked out to the area where Harry Walker died. The first time, I asked to be escorted by an off-duty Yellowstone park ranger in order to legitimatize my transgression on the white barrens around the geysers, which are sensitive to foot travel and downright lethal to some unfortunate people. I wanted to go up at sunset. The forest has grown back since it burned in 1988, and the two of us sat silently as the woods grew dark around us. My scalp prickled.

The second time, I walked back out there alone, just to feel what it was like in the dark that night in 1972.

The third time I went up there with my son, James. It was late June, within a day or two of the date Harry died, and again I selected sunset as the time to go. As we left the boardwalk I saw a couple of bones lying on the white clay, the ribs of a winter-killed elk calf. They interested me because by one account Harry and Phillip had used a pair of bones, not sticks, to mark the point where they left the trail to their camp. I picked them up and carried them in my hand, and when we reached the base of the timbered ridge and began climbing I discarded them, tossing them ahead of me with no particular attention. The bones landed several feet away in the blueberries, one on top of the other, in the shape of a cross.

Contrary to what William Drury and Ian Nisbet proposed, the world has never seemed random to me. I've always felt that I live in a web of complex relations, a community, as my friend Wendell Berry puts it, that includes both people and the nonhuman, in which the ties that bind, bind in all directions. Things do not just happen randomly. Nisbet, by the way, went on from his radar work at Cape Cod to do research on terns, a bird that lives on fish, and like everyone else in this story, it seems, he wrote an article—in his case on terns—for *National Geographic*. Because terns are at the top of their food chain, they accumulate toxic chemicals such as polychlorinated biphenyls, or PCBs. Through his tern studies Nisbet became an expert on the environmental effects of PCBs.

I learned this when I called him up to ask about his radar work. From there the conversation turned to PCBs, and he told me about a major lawsuit in which he testified as an expert witness. A Monsanto factory where PCBs were produced had dumped them in a creek for years. That creek flowed into another, just downstream, called Choccolocco Creek. The plant was in Anniston, Alabama, and luckily the waste reached Choccolocco Creek miles downstream from the Walker farm.

The most important outcome of the Craigheads' grizzly work and the radio collar was the connection of Yellowstone National Park, Grand Teton National Park, five national forests, and two national wildlife refuges into a single landscape. Ultimately this benefited not just the grizzly but all the other forms of life that thrive with greater landscape connectivity. "The radio collar showed us where the grizzlies went, and where the grizzlies went showed us the Greater Yellowstone Ecosystem," remarked a Montana conservationist, after Frank Craighead's death. The new Leopold Report, "Revisiting Leopold," affirmed the value of this: "Connectivity across . . . broader land- and seascapes is essential for system resilience over time." It is no longer enough to have island-like strongholds of national parks. Now, for wildness to survive it must be connected so that animals and plants can have genetic interchange and, if necessary, adjust their ranges during climate change.

Upstream from the PCB cleanup, Choccolocco Creek has become one such connected landscape. As Wallace Walker slowly went out of business following the reversal in the Court of Appeals, he sold off all but twelve acres, including his beautiful pasture. For a while, the pasture's new owners were going to sell it to a major corporation for development.

But a conservationist doctor and his wife bought five hundred acres, including Wallace's pasture, and they are now working with other landowners under the aegis of a new organization called the Choccolocco Creek Watershed Alliance to preserve the creek and the valley. Harry's niece Renee is on the board of directors. Coyotes are moving up and down the creek. Broad-winged and red-tailed hawks hunt rabbits in Wallace's pasture now. Bluebirds nest on the land. A big oak that used to shade the farmhouse fell down, and the barn owls that lived in it moved to the neighboring forest. Killdeer nest near where the barn stood. Deer, raccoons, groundhogs, and opossums tread the old cow paths. Daisy fleabane, oxeye daisy, wild petunia and violets, henbit mint, jewelweed, black-eyed Susan, and Queen Anne's lace bloom on the farm.

Frank Craighead is gone, and so is Glen Cole. I think of the Thornton Wilder play *Our Town,* in which the dead talk to one another in the town graveyard as the passions of life fall away. Surely if there exists a residue after the body dies it would have no time for life's disagreements. What united Frank and Glen was their love of wildness. Wildness was why Harry grew a beard, why he called home to have Wallace send his camping gear to Wyoming, and why he was singing about a bear at the time he was killed by one.

William Drury died in 1992. A book of his essays, including his attack on the theory of succession, was published posthumously. Ian Nisbet revered him and has never recanted his belief in the fundamental randomness of nature. As I wrote, I am a dissenter on that issue. Everything seems too well organized to me to be a mere appearance of organization that is really the result of a series of accidents. And if order prevails in nature, then it prevails in my life, and yours. And if this is true, then the beautiful world—the care of which incited such bitter argument between the people in this book—has a purpose, and so did the life of Harry Walker, and so do ours.

ACKNOWLEDGMENTS

ONE EVENING IN the autumn of 1998, I brought a fifth of single-malt Scotch, a notebook, and a tape recorder to the home of David Graber to record his recollections of Starker Leopold. This began a long series of conversations with Graber about ecological intervention and restoration in national parks and wilderness. Later, when I needed to find someone who was really against such manipulation, I sought out the historian Roderick Nash, and we made a series of trips by white-water raft and oceangoing yacht to look at examples of what he derisively called, after Howard Zahniser, "gardening." Both men were incredibly generous with their time, advice, and insight, and the most valuable product, to me, of the time we spent together, was not this book, but the pleasure of their friendship.

I was supported all the way through the journey by my agent, Sandra Dijkstra, and her staff—Elise Capron, Elizabeth James, and Andrea Cavallaro—who have made my life as a writer possible. I will always be grateful to John Glusman for acquiring this book for Penguin Random House. I received advice early in the writing process from his successor, Charlie Conrad. It fell to senior editor Kevin Doughten to edit the book from start to finish, and I was extremely lucky in that regard. Kevin's unerring judgment, his capacity for hard work, his wit, and the sheer pleasantness of knowing him have endeared him to me.

A very special thank-you is due to Crown senior production editor Terry Deal, without whose hard work and endless patience with my tendency to improve things this book could not have gone to print. Big thanks, Terry! I would also like to recognize my publicist, Liz Esman; my serial rights specialist, Courtney Snyder; and my marketer, Sarah Pekdemir. Also Jeffery Ward, for his careful work on the book's maps.

John Craighead Jr., Lance Craighead, Charlie Craighead, and Karen Haynam were generous in sharing their recollections as well as access to their fathers' homes, papers, photographs, and effects—what a thrill it was to handle the first radio collars. I am eternally grateful for their kindness. Professor James Peek of the University of Idaho knew and rubbed shoulders with the Craigheads, Glen Cole, and the other scientists in this book. From the time I first made his acquaintance at a Wildlife Society meeting, he patiently explained the people and concepts of wildlife ecology from the 1950s to the '80s. Over a period of years he gave me what amounted to a graduate seminar in the history of science. His review of sections of the manuscript saved me from my ignorance. Former chief research biologist Dr. Mary Meagher shared insights gathered over her half-century of experience at Yellowstone, as well as her recollections of her graduate advisor, Starker Leopold. She was incredibly patient and generous with her time. She is regarded as a national treasure by those who worked with her and as a fearless pioneer by the many women who followed her into Park Service careers. Dr. Meagher and Dr. Douglas Houston showed grace and humility in discussing with me the history of natural regulation. No scientist who moves beyond what is safely known into what is not yet known—which is where science is made—has ever been right all the time. In the historic realization of Starker's dream of deploying teams of PhDs into the national parks, these two great people made important contributions to the protection of nature and the advancement of knowledge.

The staff of the Yellowstone Heritage and Research Center—the historian Lee Whittlesey, chief archivist Colleen Curry, archivists Anne Foster and Debra Guernsey, and librarians Jackie Jerla nad Jessica Gerdes—served me with such care and pride and love for the priceless records they curate of the first national park in the world. The reference librarian Roberto Delgadillo of the University of California, Davis, has been irreplaceable in assisting my search for obscure things and kept me laughing with his madcap humor.

Everything I learned about wildfire in the course of my work on this book was contextualized in the conversations I had with the fire ecologist Kevin Robertson at Tall Timbers Research Station, in Florida, including those we had on a canoeing trip in the Okefenokee Swamp, during which we collided with at least three submerged alligators. Thank you, Kevin, for your tutelage and friendship.

Harry Walker's sisters and their husbands, Betty and J. C. Simmons, Carolyn and Walter Crowe, and Jenny and Eddie Whitman, and Harry's niece, Renee Simmons Raney, welcomed me to their farm and shared their lives and memories of Harry. They made me feel like family myself. Jim Brady was indefatigable in revealing to me the world of Park Service rangers in the 1960s and '70s, and he generously assisted me in contacting his countless friends and admirers from the national parks. Nathan Stephenson, the Sequoia and Kings Canyon National Parks forest ecologist, offered me his friendship and answered my endless questions about his field and science in the national parks.

Charles Zetterberg graciously assisted me in learning about his father and giving me access to his papers.

For their friendship, encouragement, and support over the years, I would like to thank Joanna Robinson, Keith Hadley, David Lukas, Sonika Tinker, Christian Pedersen, Brett and Louis Jones, Amy Tan and Louis DeMattei, Wendell and Tanya Berry, Terry Tempest and Brooke Williams, Gary Snyder, Barry Lopez and Debra Gwartney, Susan Nance, Kara deVries, and Melissa Seibold. For more of the generous people without whom this work could not have been done, see the Notes section, beginning on page 334.

NOTES

1: Los Angeles

The transcripts of *Dennis G. Martin, as Administrator of the Estate of Harry Eugene Walker, Deceased, Plaintiff, v. United States of America, Defendant* are in the Yellowstone National Park Heritage and Research Center. Courtroom testimony can be halting and at times it diverts into arguments between lawyers over technical matters that are incomprehensible to even the most well-informed layperson. For the benefit of the reader, in places, I have edited the excerpts from testimony in *Martin v. United States* included in this book, but I haven't changed the original words or added ones that were not present in the transcripts. Other legal records associated with the case are in the National Archives at Riverside, California. My thanks to Kerry Bartels there.

The papers of Stephen Zetterberg and the memories of his son, Charles, were invaluable in reconstructing Zetterberg's role. The recollections of Dennis Martin and the former Interior Department solicitor Curt Menefee were also key in bringing the trial to life. Judges Larry Crispo and Thomas Smith and Justices Kathleen Butz and William Rylaarsdam of the California Courts of Appeal explained legal concepts and procedures. My research on the character of William Spivak was assisted by his cousin Dianne Shiner. My understanding of Judge Andrew Hauk was informed by interviews with judges and attorneys who knew him, including Dennis Martin and Larry Crispo. Judge Hauk was also profiled in a less-than-flattering account in Joseph C. Goulden's *The Benchwarmers: The Private World of the Powerful Federal Judges* (New York: Weybright and Talley, 1974) and in coverage of his many important cases in the *Los Angeles Times* and by the wire services.

2: American Eden

My thanks to Dr. Jean Merriman and Stephen Studebaker, both seasonal rangers for the National Park Service (now retired), for introducing me to

Yellowstone. Yellowstone is a publishing industry in and of itself, and I will list here only a few of the dozens of sources I reviewed in the course of writing this book. The National Park Service publishes annually its spiral-bound compendium of up-to-date natural and cultural history and administrative issues, *Yellowstone Resources and Issues Handbook: 2015* (Yellowstone National Park, WY: Yellowstone National Park, 2015). Aubrey L. Haines's two-volume *The Yellowstone Story: A History of Our First National Park*, rev. ed. (Boulder: University Press of Colorado, 1996–99), is the definitive work on Yellowstone's history. Paul Schullery's *Searching for Yellowstone: Ecology and Wonder in the Last Wilderness* (Helena: Montana Historical Society Press, 2004) is an excellent companion to Haines. Schullery and his sometime coauthor the Park Service historian Lee Whittlesey are the foremost living experts on Yellowstone history. Both of them have been generous with their time in answering my questions. Anne G. Harris, Esther Tuttle, and Sherwood Tuttle, *Geology of National Parks* (Dubuque: Kendall-Hunt Publishing, 1995) was helpful in describing Yellowstone's geography, as was Robert L. Taylor, Joseph M. Ashley, and William W. Locke, III, *Geological Map of Yellowstone National Park* (Bozeman: Montana State University, 1989). Joel Achenbach, "When Yellowstone Explodes" (*National Geographic*, August 2009, 56–69), is an excellent account of Yellowstone's volcanism and was the source for the story about Lieutenant Doane's 1870 observation of the Yellowstone Caldera. Paul K. Doss and Amy Bleichroth, "Following the Path of Stone: Obsidian Artifacts from Indiana Sourced to Yellowstone Plateau" (*Yellowstone Science* 20, no. 2 [September 2012]: 12–14), covers recent research on obsidian trade routes.

3: YOSEMITE AND YELLOWSTONE

Harold K. Steen's *The U.S. Forest Service: A History* (Seattle: University of Washington Press, 1976) gives a good account of the nineteenth-century context for the emergence of the conservation movement as well as a nice organizational history of the agency.

For a general history of Yosemite National Park, see Alfred Runte, *Yosemite: The Embattled Wilderness* (Lincoln: University of Nebraska Press, 1990). The National Park Service's excellent online library also includes this book at www.nps.gov/parkhistory/online_books/rusticarch/contents.htm.

Joseph H. Engbeck Jr. has also written an excellent treatment of Yosemite's pre-national-park years as a California state park in chapter 1 of *State Parks of California: From 1864 to the Present* (Portland, OR: C. H. Belding, 1980).

For a treatment of Eden as a symbol in national parks, see Mark Stoll, "Milton in Yosemite: *Paradise Lost* and the National Parks Idea" (*Environmental History* 13, no. 2 [2008]: 237–74).

The Yosemite Online Library, a private website not affiliated with the National Park Service, has a wealth of Yosemite-ana with html links, including many out-of-print and archival materials. I found it extremely useful: www.yosemite.ca.us/library/.

The National Park Service also has a library and an archive at Yosemite. I wish to thank librarian Jackie Zak for her help there.

4: APPALACHIAN SPRING

The Walker family's patience with and good humor through more than a hundred e-mails, many hours of interviews, and who knows how many telephone calls astounded me. Their photographs, home movies, and letters, and Jenny's journal entries, were essential to bringing the farm in the 1960s and '70s to life. Renee Simmons Raney, Harry's niece and an expert naturalist, was extremely helpful in assisting me with the natural and cultural history of the Walker farm. Renee is the assistant director for the Jackson State University Environmental Policy and Information Center and Field Schools, director of conservation for the Georgia-Alabama Land Trust, and the 2014 recipient of the Roosevelt-Ashe Conservation Award for Outstanding Educator in Conservation. Also helpful were the Anniston Museum of Natural History and the Public Library of Anniston-Calhoun County's historical collection. I would also like to thank W. Pete Conroy of Jackson State University, Agnes Owen "Wen" Scherer, Earl Walden, and former *Anniston Star* journalist Dennis Love.

5: FRANK

In addition to interviews with members of their families, John and Frank Craighead's own writings and papers were essential to my understanding their lives and characters. They include John J. Craighead and Frank C. Craighead Jr., *Life with an Indian Prince* (Surrey, Canada: Hancock House, 2001); Frank C. Craighead Jr., *A Naturalist's Guide to Grand Teton and Yellowstone National Parks* (Guilford, CT: FalconGuides, 2006); Frank and John Craighead, *Hawks in the Hand: Adventures in Photography and Falconry* (Boston: Houghton Mifflin, 1939), and *Hawks, Owls and Wildlife* (Harrisburg, PA: Stackpole, 1956); John and Frank Craighead, "Adventures with Birds of Prey" (*National Geographic* [July 1937]: 109–34), "In Quest of the Golden Eagle" (*National Geographic* [May 1940]: 693–710), "We Survive on a Pacific Atoll" (*National Geographic* [January 1948]: 73–94); and Frank and John Craighead, "Cloud Gardens in the Tetons" (*National Geographic* [June 1948]: 811–30), "Wildlife Adventuring in Jackson Hole" (*National Geographic* [January 1956]: 1–36), and "Bright Dyes Reveal Secrets of Canada Geese" (*National Geographic* [December 1957]: 817–32).

6: THE BALANCE OF NATURE

For an excellent history of the science of ecology, see Donald Worster, *Nature's Economy: A History of Ecological Ideas,* 2nd ed. (New York: Cambridge University Press, 1994), and *The Wealth of Nature: Environmental History and the Ecological Imagination* (New York: Oxford University Press, 1993). The history of the management of nature in national parks is covered in Richard West Sellars, *Preserving Nature in the National Parks: A History* (New Haven, CT: Yale University Press, 1997), and James A. Pritchard, *Preserving Yellowstone's Natural Conditions: Science and the Perception of Nature* (Lincoln: University of Nebraska Press, 1999).

For an account of the Ecological Society of America's program to preserve natural conditions in parks, see Victor E. Shelford, "Twenty-Five-Year Effort at Saving Nature for Scientific Purposes" (*Science,* September 24, 1943, 280–81), and "Nature Sanctuaries" (*Science,* May 6, 1932, 481), and Gina Rumore, "Preservation for Science: The Ecological Society of America and the Campaign for Glacier Bay National Monument" (*Journal of the History of Biology* 45, no. 4 [November 2012]: 613–50, published electronically September 29, 2011).

With regard to fires in Glacier National Park as the driver of fire organization in the Park Service, see Hal K. Rothman, *Blazing Heritage: A History of Wildland Fire in the National Parks* (New York: Oxford University Press, 2007).

Charles Adam's 1925 article sounding the alarm on the decline of natural conditions in national parks is "Ecological Conditions in National Forests and in National Parks" (*The Scientific Monthly* 20, no. 6 [June 1925]: 561–93).

7: BERKELEY

The story of Joseph Grinnell and the Museum of Vertebrate Zoology is covered in Barbara R. Stein, "Annie M. Alexander: Extraordinary Patron" (*Journal of the History of Biology* 30, no. 2 [Summer 1997]: 243–66), E. R. Hall, "Joseph Grinnell (1877 to 1939), *Journal of Mammalogy,* 20 no. 4 [1939]: 409–17, and H. W. Grinnell, "Joseph Grinnell: 1877–1939, *Condor,* 42, no. 1 [1939]: 2–34. Regarding his record of the last confirmed grizzly footprint in the San Gabriel Mountains, see Tracy I. Storer and Lloyd P. Tevis Jr., *California Grizzly* (Berkeley: University of California Press, 1983) 28. For a sense of the connection between UC Berkeley and the National Park Service, see Glen Martin, "Love National Parks? Thank UC Berkeley, and What Transpired Here 100 Years Ago" (*California,* March 19, 2015, http://alumni.berkeley.edu/california-magazine/just-in/2015-03-23/love-national-parks-thank-uc-berkeley-and-what-transpired).

See also Sellars, *Preserving Nature in the National Parks.*

My thanks to Pamela Wright Lloyd, George Wright's daughter, for her assistance in my research about her father, George M. Wright. She and her son-in-law, Jerry Emory, published a short biography, "George Melendez Wright, 1904–1936: A Voice on the Wing" (*George Wright Forum* 17, no. 4 [2000]: 15–44). See also George Wright's own publications in *The Condor,* as well as his two survey reports on wildlife conditions in the national parks: George M. Wright, Joseph S. Dixon, and Ben H. Thompson, *Fauna of the National Parks of the United States: A Preliminary Survey of Faunal Relations in National Parks* (Washington, DC: US Government Printing Office, 1933), and George M. Wright and Ben H. Thompson, *Fauna of the National Parks of the United States: Wildlife Management in the National Parks* (Washington, DC: US Government Printing Office, 1935).

8: CLAYPOOL

My questions about Yellowstone bear history and management were patiently answered by Kerry A. Gunther, Yellowstone's excellent bear biologist for over three decades. My grateful thanks to Kerry, who always made time to see me when I was around.

Lee H. Whittlesey, the National Park Service's resident Yellowstone historian, has written an exhaustive treatment of deaths by grizzly attack (and by every other means): *Death in Yellowstone: Accidents and Foolhardiness in the First National Park,* 2nd ed. (Lanham, MD: Roberts Rinehart, 2014). For his account of the death of Martha Hansen, see pages 60–61.

For the lawsuit over William Claypool's injuries at Old Faithful, see *Claypool v. United States* (98 F. Supp. 702 [S.D. Cal. 1951]), as well as contemporaneous newspaper accounts. See also Paul Schullery, *The Bears of Yellowstone* (Worland, WY: High Plains, 1992).

See also Mark A. Haroldson, Charles C. Schwartz, and Kerry A. Gunther, "Grizzly Bears in the Greater Yellowstone Ecosystem: From Garbage, Controversy, and Decline to Recovery" (*Yellowstone Science* 16, no. 2 [2008]: 13–23); Mary Meagher, "Bears in Transition, 1959–1970s" (*Yellowstone Science* 16, no. 2 [2008]: 5–12); Kerry A. Gunther, "Bear Management in Yellowstone National Park, 1960–93 (*Bears: Their Biology and Management* 9, pt. 1: *A Selection of Papers from the Ninth International Conference on Bear Research and Management,* Missoula, MT, February 23–28, 1992 [1994]: 549–60).

The Craigheads' first year and a half of work in Yellowstone is described in Frank and John Craighead's "Knocking Out Grizzlies for Their Own Good" (*National Geographic,* August 1960, 276–91), and in Frank's memoir, *Track of the Grizzly* (San Francisco: Sierra Club Books, 1979). John J. Craighead, Jay S. Sumner, and John A. Mitchell's *The Grizzly Bears of Yellowstone: Their Ecology in the Yellowstone Ecosystem, 1959–1992* (Washington, DC: Island Press, 1995) covers the Yellowstone work from start to finish. I have made extensive use of the correspondence and papers of the Craigheads, of National Park staff, and of Starker Leopold.

My thanks to Maurice Hornocker for his interviews and for checking in manuscript the accounts he gave me for accuracy and completeness, as well as to the former Craighead assistants Jay Sumner and Joel Varney, Lance Craighead, and John Craighead Jr. for allowing me to interview them.

9: Smitty

In all maps, reports, and other references from 1960, the path on which Smitty's attack took place was referred to as Roes Creek Trail. Today, it is known as the Rose Creek Trail. To prepare this account of what happened there I used the records of *Parratt v. United States* (Central District of California, Civil No. 64-435-JWC), as well as my fieldwork in Glacier National Park. The late Alan Nelson, who was mauled in the same attack, assisted me greatly in our interviews and in his review of the manuscript before his death. I also wish to thank Smitty's brother Mark W. Parratt, whose correspondence and willingness to be repeatedly interviewed I greatly appreciated. Mark's memoir of his family's lives at Glacier National Park and of Smitty's accident, *Fate Is a Mountain* (Whitefish, MT: Sun Point Press, 2009), was also very helpful, as was an account in Albert Ruffin, "Attacked by a Grizzly Bear: The Ordeal of a Boy All but Eaten Alive" (*Life,* August 27, 1965, 73–82).

Steve J. Gniadek and Katherine C. Kendall's "A Summary of Bear Management in Glacier National Park, Montana, 1960–1994" (*Ursus,* no. 10 [1998]: 155–59), and Clifford J. Martinka's "Preserving the Natural Status of Grizzlies in Glacier National Park" (*Wildlife Society Bulletin* 2, no. 1 [Spring 1974]: 13–17) were also useful.

10: Trout Creek

Ian Nisbet was generous with his time in explaining the history of radar work on birds. An unpublished interview with him about his life and career was also useful, as was Ted R. Anderson's *The Life of David Lack: Father of Evolutionary Ecology* (New York: Oxford University Press, 2013).

For a review of injuries caused by and incidents involving bears in Yellowstone, see Kerry Gunther, "Bear Management in Yellowstone National Park, 1960–93" (*Bears: Their Biology and Management* 9, no. 1 [1994]: 549–60).

See also Mary Meagher, "Bears in Transition: 1959–1970s" (*Yellowstone Science* 16, no. 2 [2008]: 5–12) and Mark A. Haroldson, Charles C. Schwartz, and Kerry A. Gunther, "Grizzly Bears in the Greater Yellowstone Ecosystem: From Garbage, Controversy, and Decline to Recovery" (*Yellowstone Science* 16, no. 2 [2008]: 13–23).

I made extensive use of Etienne Benson's *Wired Wilderness: Technologies of Tracking and the Making of Modern Wildlife* (Baltimore: Johns Hopkins University Press, 2010), a truly fine contribution to the history of science that is about the Cold War–era adoption of surveillance technologies by wildlife biologists. See also Craighead, Sumner, and Mitchell, *Grizzly Bears of Yellowstone;* Pritchard, *Preserving Yellowstone's Natural Conditions;* and Schullery, *Bears of Yellowstone* and *Searching for Yellowstone.*

11: THE BIG KILL

My special thanks to Maurice Hornocker for helping me render the scene of his and John Craighead's visit to the elk-herd reduction in the Lamar Valley in January of 1962. Photos of elk reduction operations in the collections of the Yellowstone Heritage and Research Center were essential to getting a feel for what went on there, as was my interview with John Good, who participated in the shooting. Special thanks also to Bill Ripple at Oregon State University for our many conversations about elk and trophic cascades.

A detailed eleven-page report on the Big Kill by Robert Howe, Yellowstone's management biologist at the time, was published as "Final Reduction Report, 1961–62: Northern Yellowstone Elk Herd" (US National Park Service, 1962), and the report that preceded it—which is really a sort of prospectus, something like an environmental impact statement to build public and agency support for what was about to happen—is "Management of Northern Yellowstone Wildlife and Range" (November 10, 1962). "Yellowstone River and Gallatin River Elk Herds: Management Progress and Problems" (US Forest Service, Northern Region, 1961), a report by the Forest Service's assistant regional forester, W. W. Dresskill, contains the minutes of a June 1961 public meeting with representatives from the Park Service, Forest Service, and State of Montana about the coming elk reduction. The Montana Fish and Game Commission, representing hunters, obviously doesn't want the killing to happen. However, the mere fact of all parties being present—which at that point had already been going on for years in the form of the Absaroka Conservation Committee's interagency coordination on managing the migratory northern elk—serves as a precursor to the Greater Yellowstone Ecosystem's cross-boundary management process that would come later.

A special issue of *Yellowstone Science,* "Ungulate Management in Yellowstone, Part II: Oral History Interviews with Former Staff" (*Yellowstone Science* 8, no. 2 [Spring 2000]: Yellowstone Association for Natural Science, History, and Education, 2000), contains interviews with Glen Cole and John Good.

A characteristically excellent article produced under Bruce Kilgore during his tenure as editor of the *Sierra Club Bulletin,* "Is Hunting in the National Parks the Answer for Too Many Elk?" (*Sierra Club Bulletin* 47, no. 8

[November 1962]: 4–5, 9), gives a feel for the controversy that followed the herd reduction of 1962.

For a fascinating history of "trophic cascades" and of the outcomes of the eradication of predators, see William Stolzenburg, *Where the Wild Things Were: Life, Death, and Ecological Wreckage in a Land of Vanishing Predators* (New York: Bloomsbury, 2008).

The figure for the numbers of predators hunted down in Yellowstone between 1904 and 1935 is from Adolph Murie's *Ecology of the Coyote in the Yellowstone: Fauna Series No. 4* (US National Park Service, 1940).

Under George Bird Grinnell and others, *Forest and Stream* ran at least thirteen articles between 1880 and 1910 about the plight of elk and other big game at Yellowstone. These give insight into Grinnell's efforts to stop poaching at Yellowstone, as well as his concerns about winter die-offs, which led to the winter feeding of elk and other herbivores and the creation of the National Elk Refuge. Frederic H. Wagner's *Yellowstone's Destabilized Ecosystem: Elk Effects, Science, and Policy Conflict* (New York: Oxford University Press, 2006) contains elk censuses as well as detailed descriptions of elk's effect on plant life in the Northern Range.

Historical surveys of the Northern Range situation are also given in Wright, Dixon, and Thompson, *Fauna of the National Parks*; Gerald R. Wright, "Wildlife Management in the National Parks: Questions in Search of Answers" (*Ecological Applications* 9, no. 1 [February 1999]: 30–36); and Frederic H. Wagner et al., *Wildlife Policies in the U.S. National Parks* (Washington, DC: Island Press, 1995).

12: STARKER

I am deeply grateful to Starker Leopold's sisters, Estella Leopold and the late Nina Leopold Bradley, for sharing their recollections with me. Nina had written me an encouraging letter in 1996 in response to my first published essay. To get an admiring letter from her went to my head to such an extent that it kept me writing for years. I will never forget my visits with Nina or the warmth Estella has shown in our many phone calls and e-mails. My two visits to the Shack and my solitary ramblings on the Leopolds' Sauk County farm—the second time on Nina's late husband's old wooden cross-country skis—were spiritual experiences.

The Leopolds were excellent at chronicling their own lives in journals and photographs, and the University of Wisconsin has one of the finest online archives of such materials I have ever seen. It is highly recommended. I'm a huge admirer of Curt Meine's stunning biography *Aldo Leopold: His Life and Work* (Madison: University of Wisconsin Press, 2010), which gives insight into Leopold's family life and activities with Starker. The best introduction to Aldo Leopold's ideas—and those of Starker, who followed him—

are his own essays, collected in *A Sand County Almanac: And Sketches Here and There* (New York: Oxford University Press, 1949), *Round River: From the Journals of Aldo Leopold* (New York: Oxford University Press, 1953), and Susan L. Flader and J. Baird Callicott, eds., *The River of the Mother of God: And Other Essays* by Aldo Leopold (Madison: University of Wisconsin Press, 1991). The Leopold scholar Susan L. Flader's own *Thinking like a Mountain: Aldo Leopold and the Evolution of an Ecological Attitude Toward Deer, Wolves, and Forests* (Madison: University of Wisconsin Press, 1994) is an excellent history of Aldo's thought as well. Thanks to Susan for our correspondence and meetings.

I made extensive use of Starker Leopold's papers at the Bancroft Library at UC Berkeley, as well as the then-unaccessioned ones that were still in the basement of Mulford Hall but have since been added by the Bancroft. I will never forget my days reading the latter papers, alone in Leopold's office next to the wall of books on wildlife and natural sciences that he had inherited from his father. Thanks to Professor Reg Barrett for trusting me with that privilege, and to Professor Dale McCullough and others of Leopold's coworkers for their recollections.

Carol Henrietta Leigh Rydell's 1993 master's thesis, "Aldo Starker Leopold: Wildlife Biologist and Public Policy Maker," is excellent (Montana State University, Bozeman, 1993). The 1983 retrospective interview with Starker Leopold by Carol Holleufer for the Sierra Club Oral History Project in the collection of UC Berkeley's Bancroft Library was also invaluable. Leopold's own articles and publications were useful.

The history of ecological restoration is well covered in William R. Jordan III and George M. Lubick's *Making Nature Whole: A History of Ecological Restoration* (Washington, DC: Island Press, 2011), which also covers wetland restoration projects undertaken by the same San Jose State scientists who did the pioneering work on fire in Sequoia and Kings Canyon National Parks.

The story of the University of Wisconsin–Madison Arboretum is exhaustively covered in Franklin E. Court, *Pioneers of Ecological Restoration: The People and Legacy of the University of Wisconsin Arboretum* (Madison: University of Wisconsin Press, 2012). The University of Wisconsin has put up selections, including Aldo Leopold's speech at the dedication ceremony, from William R. Jordan III, ed., *Our First 50 Years: The University of Wisconsin Arboretum, 1934–1984* (Madison: University of Wisconsin Arboretum, 1984), at http://digital.library.wisc.edu/1711.dl/EcoNatRes.ArbFirstYrs.

Aldo and Starker Leopold's hunting trip to the Sierra Madre is covered in their own meticulous journal entries, including pasted-in photos, collected in the Aldo Leopold Archives at the University of Wisconsin and available online. I have described Aldo Leopold's death from Curt Meine's account of it.

Many thanks to Curt and Susan Flader for their friendliness and encouragement.

13: PROMETHEUS

The chart of the Museum of Vertebrate Zoology's academic lineage, with Joseph Grinnell alone at the root, is in the history section of the museum's website at http://mvz.berkeley.edu/Grinnell_Lineage.html.

Grateful thanks to Bruce Kilgore, Howard Shellhammer, Tony Caprio, and Nathan Stephenson for taking a bunch of us onto Redwood Mountain and into Redwood Canyon in June of 2008 to show us the sites of the early burns, during which Kilgore located one of his calorimeters—it was really just a coffee can full of water; the fire's energy release could be calculated from how much water boiled off. Bruce Kilgore's correspondence, which I found liberally sprinkled in the papers of others, shows the energy and focus with which he carried out Starker Leopold's vision of bringing fire back to the national parks. His "Origin and History of Wildland Fire Use in the U.S. National Park System" (*George Wright Forum* 24, no. 3 [2007]: 92–122) is a great introduction to the setting of fires from someone who witnessed—no, who did—it.

After all these years in print, Stephen Pyne's towering achievement *Fire in America: A Cultural History of Wildland and Rural Fire* (Seattle: University of Washington Press, 1997) is still the most comprehensive history of the subject, and it's also really easy to read. My thanks to Stephen for our phone calls and correspondence, as well as to Arizona dendrochronologist Tom Swetnam.

Hal Rothman's *Blazing Heritage* is the authoritative work on how the reintegration of fire in the West was made possible by tests in national parks. Ashley L. Schiff's *Fire and Water: Scientific Heresy in the Forest Service* (Cambridge: Harvard University Press, 1962) is the best treatment of what I call the fire insurgencies in the Southeast and West.

Timothy Egan's *The Big Burn: Teddy Roosevelt and the Fire That Saved America* (Boston: Houghton Mifflin Harcourt, 2009) is a good history of the 1910 conflagrations in Montana and Idaho that propelled the Forest Service to become "the nation's fire department."

Herbert Stoddard's *Memoirs of a Naturalist* (Norman: University of Oklahoma Press, 1969) is a look at a fairly recent time period when a bright and extremely hardworking person could rise to the top of a scientific profession without ever having gone to college. See also Albert G. Way, "Burned to Be Wild: Herbert Stoddard and the Roots of Ecological Conservation in the Southern Longleaf Pine Forest" (*Environmental History* 11, no 3 [July 2006]: 500–26). Deepest thanks to my friend Kevin Ferguson, fire ecologist at Tall Timbers Research Station, for his gift of the station's history in Robert L. Crawford and William R. Brueckheimer's *The Legacy of a Red Hills Hunting Plantation: Tall Timbers Research Station and Land Conservancy* (Gainesville: University Press of Florida, 2012).

See also H. H. Biswell's "Research in Wildland Fire Ecology in California" in Roy Komarek, ed., *Proceedings Second Annual Tall Timbers Fire Ecology Conference* (Tallahassee: Tall Timbers Research Station, 63–97).

Starker Leopold's speech at the 1957 Wilderness Conference was published in Sierra Club, *Fifth Biennial Wilderness Conference, Summary of Proceedings* (San Francisco: Sierra Club, 1957). His story about the shouting match with Harold Bryant was in the Carol Holleufer interview, as was his account of Secretary Udall's initial response to the Leopold Report.

14: Observable Artificiality in Any Form

The National Park Service has the full text of the Leopold Report online at www.nps.gov/parkhistory/online_books/leopold/leopold3.htm.

For assessments of what the report meant at the time, see Sellars, *Preserving Nature in the National Parks,* and Pritchard, *Preserving Yellowstone's Natural Conditions.*

Starker Leopold shared his own recollections of and perspective on the report in an address to a meeting of park superintendents, reprinted with a preface by David Graber in "Starker Leopold's Second Thoughts on the Leopold Report: A Recently Discovered Transcript of a 1975 Speech" (*George Wright Forum* 30, no. 2 [2013]: 200–11.

15: Reconstruction

The Craigheads' research on Yellowstone elk was published in two bound Wildlife Society monographs: John J. Craighead, Gerry Atwell, and Bart O'Gara, *Elk Migrations in and Near Yellowstone National Park* (*Wildlife Monographs,* no. 29; The Wildlife Society, Washington, DC, 1972), and John J. Craighead, Frank C. Craighead, Robert L. Ruff, and Bart W. O'Gara, *Home Ranges and Activity Patterns of Nonmigratory Elk of the Madison Drainage Herd as Determined by Biotelemetry* (*Wildlife Monographs,* no. 33; The Wildlife Society, Washington, DC, 1973). I also referred to John Craighead's notes.

The progress of the Smitty Parratt case was reconstructed from Stephen Zetterberg's papers, court records, and the recollections of Alan Nelson and Mark and Kay Dell Parratt.

The San Jose State sequoia researchers Richard J. Hartesveldt, Tom Harvey, Howard Shellhammer, and Ronald Stecker's work was published in *Giant Sequoia Ecology: Fire and Reproduction* (Scientific Monograph Series, no. 12; Washington, DC: US Department of the Interior, National Park Service, 1980). It's available in full at www.nps.gov/parkhistory/online_books/science/12/contents.htm.

Jan van Wagtendonk, who did his doctoral work under Harold Biswell, surveyed Biswell's role in "Dr. Biswell's Influence on the Development of Prescribed Burning in California," in the proceedings published as *The Biswell Symposium: Fire Issues and Solutions in Urban Interface and Wildland Ecosystems* (USDA Forest Service General Technical Report PSW-GTR-158, 1995). Contemporaneous profiles of Biswell's work in the Whitaker Forest are Geraldine B. Larson, "Whitaker's Forest" (*American Forests* 72 [September 1966]: 22–25, 40); Harold Weaver and Harold Biswell, "Redwood Mountain" (*American Forests* 74 [1968]: 20–23); and a reprint, H. H. Biswell, R. P. Gibbens, and Hayle Buchanan, "Fuel Conditions and Fire Hazard Reduction Costs in a Giant Sequoia Forest" (*National Parks Magazine* 42, August 1968, 17–19), in the Sequoia and Kings Canyon National Parks archives. My thanks to the archivist Ward Eldredge for his assistance during my visits there.

The records of the Vaughan case are in the Yellowstone National Park Archives.

The Craigheads' 1966 *National Geographic* article is "Trailing Yellowstone's Grizzlies by Radio" (*National Geographic,* August 1966, 252–67). I was able to find a copy of their 1967 National Geographic Television special, "Grizzly," in the Pacific Film Archive. It features the scene in which the grizzly Ivan the Terrible comes to during handling and chases the brothers and their assistants to their car, then jumps up on the hood and attacks the vehicle as they are careening through the sagebrush at the Trout Creek dump.

The names of Anniston and Oxford war dead are from veterans' organizations, the *Anniston Star,* and the recollections of Harry's older sisters. My thanks to Lieutenant Colonel Butler Green (ret.), the National Guard commander and state milk inspector who recruited Harry into the National Guard, for my interview with him in Anniston.

I found the 1967 television footage of rangers and butchers cutting the legs off freshly killed elk and dragging them to the processing area behind Sno-Cats still filed away at the CBS News archives, at 524 West Fifty-seventh Street in Manhattan.

The proceedings, exhibits, letters of comment, and transcripts of Senator McGee's hearings in Casper, Wyoming, were published in US Department of the Interior, "Senate Hearings: Control of Elk Population, Yellowstone National Park" (90th Congress, First Session, Fiscal Year 1968; Washington, DC: US Government Printing Office, 1967).

16: COLE

The recollections of Gladys Cole, Glen's daughter Patty Stenske, Doug Houston, Mary Meagher, Dick Mackie, and Jim Peek were invaluable to understanding the character, personality, and thought of Glen Cole. I was struck by the affection they all had for him. Cole's interview, and that of John Good, in the spring 2000 issue of *Yellowstone Science* (ibid.), were also very useful. Cole's extensive correspondence in the Yellowstone Heritage and Research Center, in Starker Leopold's papers, and in the Craighead papers also helped.

Paul Errington's classic work on the role of predators is "Predation and Vertebrate Populations," *The Quarterly Review of Biology* 21, no. 2 [June 1946]: 114–77. See also Robert E. Kohler, "Paul Errington, Aldo Leopold, and Wildlife Ecology: Residential Science," *Historical Studies in the Natural Sciences* 41, no. 2 [Spring 2011]: 216–54.

Details of the Craigheads' activities in 1967, and of various meetings such as the one with John McLaughlin about the script for "Grizzly," were from the papers of John and Frank Craighead.

17: The Night of the Grizzlies

Details of the deaths of Julie Helgeson and Michele Koons and of the circumstances surrounding them are from my fieldwork in Glacier National Park along with the Park Service's original investigative reports, statements, and photographs obtained under a Freedom of Information Act request and Jack Olsen's *Sports Illustrated* articles and book *Night of the Grizzlies.* A documentary by MontanaPBS, "Montana's Night of the Grizzlies," containing interviews with witnesses was also useful. My thanks to Denise Germann of the National Park Service for tolerating my addition to her workload, and to the Koons and Helgeson families.

18: Natural Control

My account of the meeting between John McLaughlin, Starker Leopold, Bruce Kilgore, and Forest Service fire experts from the Pacific Southwest Experiment Station is taken from Bruce Kilgore's rendition in "Origin and History of Wildland Fire Use in the US National Park System," *The George Wright Forum,* vol. 24, no. 3 (2007). Starker's exclamation to the effect that the Park Service is not asking for advice on whether to burn, but only on how, came from Kilgore's meeting notes. The emphatic italics are as Kilgore recorded them.

In my depiction of Glen Cole, I have again relied on and am grateful for the help of Gladys Cole, Patty Stenske, Doug Houston, and Dick Mackie.

Charles Kay's doctoral dissertation, "Yellowstone's Northern Elk Herd: A Critical Evaluation of the "Natural Regulation" Paradigm" (Utah State University, 1990) contains a very useful account of the development of Cole's thought from his "natural control" of 1967 to its mature form in "natural regulation" by 1971, with telling quotes from Cole's and the park's policy memoranda. I also referred to Cole's correspondence in the Yellowstone Heritage and Research Center, in the A. Starker Leopold Papers, and his scientific publications.

19: BAD BLOOD

In addition to the correspondence of the various parties, interviews with living witnesses, and the testimony in the transcripts of *Martin v. United States,* I found the historian Alice Wondrak Biel's doctoral dissertation, published as *Do (Not) Feed the Bears: The Fitful History of Wildlife and Tourists in Yellowstone* (Lawrence: University of Kansas Press, 2006), very useful in understanding the chronology of the collision between the Craigheads, Glen Cole, Jack Anderson, and Starker Leopold. More than that, in addition to being eminently readable, Biel's work is the best history of the bear situation at Yellowstone in print.

In 1988, John Craighead; Ken Greer, of the Montana Fish and Game Department; Richard Knight, the first head of the Interagency Grizzly Bear Study Team; and Helga Ihsle Pac constructed an exhaustive list of every known grizzly lost as a result of human actions before, during, and after dump closure, in *Grizzly Bear Mortalities in the Yellowstone Ecosystem, 1959–1987* (published jointly by the Montana Department of Fish, Wildlife & Parks interagency grizzly bear study team, National Fish & Wildlife Foundation, and Craighead Wildlife-Wildlands Institute, 1988). This included details such as Park Service and Craighead tag numbers, the circumstances of the death, and where it occurred. The index in Frank Craighead's *Track of the Grizzly* features an entry for each numbered bear in the text. By cross-referencing these sources with others, and where possible with narratives from witnesses to the events, I was able to detail some of the loss of the Craighead bears.

Starker Leopold's papers contained two detailed academic résumés, which along with various correspondence in his and others' papers were useful in correlating his rather overwhelming responsibilities with what was going on in his personal life and at Yellowstone. I'm grateful to Frederick Starker Leopold for his candid and compassionate reflections on his father and mother.

Bruce Kilgore detailed his fire work in extensive, carefully labeled color slides and reports. He certainly knew he was doing something that people would later be interested in. These materials are in the Sequoia and Kings Canyon National Parks archives.

20: BEAR MANAGEMENT COMMITTEE

The papers of Starker Leopold and John Craighead contain documentation of the meeting in September 1969. Frank Craighead says how unhappy he was about it in *Track of the Grizzly* (200–05). Biel, in *Do (Not) Feed the Bears,* and Pritchard, in *Preserving Yellowstone's Natural Conditions,* also give accounts. The death of the bear Marian is documented in Frank's *Track of the Grizzly* and in John Craighead et al., *Grizzly Bear Mortalities in the Yellowstone Ecosystem.*

21: FIREHOLE

Jim Brady was kind enough to share his life through extensive interviews, personal and professional papers, and photographs. Doug Houston shared his recollections of Brady, as did Tom Cherry, Scott Connelly, John Good, Mary Meagher, Rick Smith, Curt Menefee, Michael Weinblatt, and Kathy Loux. No one had anything but admiration for him. Early in our interviews, Brady casually mentioned having been shot at while working at Lake Mead in the 1960s. The three shootings are documented in Paul D. Berkowitz, *U.S. Rangers, the Law of the Land: The History of Law Enforcement in the Federal Land Management Agencies* (Redding, CA: CT, 1995). The description of Brady's flight during the elk reductions of 1967 came from beautiful color slides he took that day, as well as from his own account.

Details of the dump closures came from memoranda by Glen Cole, as well as Jim Brady's account of Rabbit Creek. Brady's recollection of the grizzly incident at his trailer matches records of a bear given to a zoo at Old Faithful in early July of 1970.

Nine-year-old Andy Hecht fell into boiling water while walking with his parents near Crested Pool at Old Faithful on June 28, 1970. Whittlesey gives the death considerable attention in *Death in Yellowstone* (19–24) because it resulted in a major campaign by the boy's father to improve safety in national parks. Jim Brady's recollections were accompanied by a photo in his own collection taken just as he pitched the bucket of soap into the water.

The Firehole River incident is extensively documented in the Walker trial testimony, as well as in the depositions taken from Tom Cherry and Jim Brady for the case. The critique of evidence collection and preservation comes from my own training and experience as a law enforcement officer and evidence custodian.

22: THE TEMPTATION OF STARKER LEOPOLD

The tendency of the Targhee National Forest (today the Caribou-Targhee National Forest) to issue sheep permits in grizzly habitat was legendary. I witnessed extensive sheep grazing in grizzly habitat there in 1978.

The forensic reconstruction of Starker Leopold's state of mind was based on his papers, his oral history, Carol Henrietta Leigh Rydell's master's thesis (ibid.), his correspondence with Cole and Anderson, two of his CVs with his work history, my discussions with his students, coworkers, and friends, and with his son Frederick Starker Leopold, to whom I am particularly indebted for his candor and obvious compassion for his late father.

The account of the deaths of bears during dump closure in this chapter and elsewhere has been reconstructed by correlating multiple sources, among them John Craighead et al., *Grizzly Bear Mortality in the Yellowstone Ecosystem,* cross-referenced by tag number with the index to individual bears in Frank Craighead, *Track of the Grizzly.*

23: Natural Regulation

The medical narrative on Harry Walker's elbow problem and subsequent surgery are from materials associated with *Martin v. United States,* in the National Archives at Riverside. For an analysis of the typical post-surgery prognosis for healing I consulted an elbow specialist at the UC Davis School of Medicine.

For more about David Lack, see Anderson, *The Life of David Lack,* and David Lack, *The Natural Regulation of Animal Numbers* (Oxford, England: The Clarendon Press, 1954). Graeme Caughley's 1970 paper is "Eruption of Ungulate Populations, with Emphasis on Himalayan Tahr in New Zealand (*Ecology* 51, no. 1 [January 1970]: 53–72).

For Charles Kay's account of how natural control in late 1967 became natural regulation in 1971, and his graph showing the precipitous drop in reports of wolf sightings, see his master's thesis, "Yellowstone's Northern Elk Herd: A Critical Evaluation of the 'Natural Regulation' Paradigm" (Utah State University, 1990).

24: Last Straws

Harry's problems with his new National Guard commander are covered in the commander's deposition and that of the company clerk in the *Martin v. United States* files at the National Archives at Riverside, and his sisters remembered some of the story.

My thanks to Gail Dormon for telling me about the Dormon boys. Evidently, the young men who presided over the Peter Pan environment of the first hippie house in the Choccolocco Valley never got old. John died young of cancer. After his death his brother Ben choked to death in a restaurant. The youngest brother, James, died young, too.

I'm grateful to the late Phillip Bradberry; his sister, Carol; his ex-wife, Marla Campbell; and his daughter, Rebecca Collins, for their interviews with me regarding Phillip's life story. His parents are long gone, but the little house on the hill above Oxford that Phillip grew up in is still there.

The mail carrier John Burdette's account of Harry's accident and his subsequent visit to the Walker farm is in his deposition, taken at Anniston in 1973. My thanks to Sharon Wilkins, the woman to whom Harry was briefly married when they were both very young, for my interview with her. For the details of what happened between the time Harry and Phillip left home and the time they met Vikki Schlicht, with the exceptions of their stay with

Harry's sister Carolyn in Louisville and Harry's calls home, we must rely on the account Phillip gave in his statement to rangers and his depositions.

25: TAKE IT EASY

It is hard to imagine the roads of America as Harry and Phillip saw them in the summer of 1972, populated as they were with thousands of hitchhikers like themselves. In a 1973 paper, "On the Road: Hitchhiking and the Highway" (*Society* 10, no. 5 [July–August 1973]: 14–21), the sociologist Abraham Miller wrote that the line of young people seeking rides at the University Avenue on-ramp to Interstate 80 in Berkeley began at dawn and stretched a mile up University Avenue. In 1974, a study by the California Highway Patrol estimated that hitchhikers were making a million trips a month in that state. There were at least two Top 40 songs about hitchhiking, the first of which was the Eagles' "Take It Easy," written by Jackson Browne and Glenn Frey.

Jackson Browne wrote "Take It Easy," but he got stuck and didn't finish it. In 1971, he gave the unfinished song to his upstairs neighbor, Glenn Frey. Frey wrote a second verse, in which the song's male protagonist is standing on a corner in Winslow, Arizona, when a young woman driving by in a Ford slows down to look him over. The hitchhiker hopes she will pick him up and that, in the words of the song, her "sweet love is gonna save me."

Frey recorded the song for the debut album of his then unknown band, the Eagles. Released that spring, "Take It Easy" was climbing the charts by June 1972 to peak at number twelve on the Top 100 during the summer migration of young hitchhikers.

I wish especially to thank Vikki Schlicht for her recollections of her brief time with Harry Walker.

26: OLD FAITHFUL

I am grateful to the Old Faithful ranger Sonja Brester for her generous hospitality, and to the ranger Colin Smith for his assistance with fieldwork.

I obtained the investigative reports on the death of Harry Walker from Yellowstone National Park through a Freedom of Information Act request. My thanks to Kerrie Evans for all the photocopying it took for the park to provide them. Some of the death scene and morgue photographs were in other files elsewhere.

The radio dispatcher's logs from the communications center at Mammoth Hot Spring for June 24 and 25, 1972, as well as the local radio log kept during the search for Harry at the Old Faithful ranger station by Elaine D'Amico, were included in the original reports. There was some question whether the food on the ground next to the tent was already stirred up to the extent it is in the scene photos when the rangers arrived, or whether some of that occurred during the search for evidence.

27: THE SEARCH FOR HARRY WALKER

Believe me when I say that from half a mile out, in the dark on the boardwalk around Geyser Hill when you're writing about a guy whose friend had just been killed by a bear, Old Faithful Village really is a tiny island of twinkling lights in a vast, dark wilderness. My thanks to the rangers of the summer of 1972 for their memories of the early morning hours of June 25, along with Vikki Schlicht, the late Phillip Bradberry, Elaine D'Amico Hall, and the Yellowstone nurse Kathy Loux.

28: MARTHA SHELL

The letters between Martha Shell and John Craighead are rich with details of Martha's tireless campaign to redeem the Craigheads and save the grizzly, and they relate to other events in news and official correspondence. Jenny Walker also provided correspondence she had with Martha. Thanks to John Craighead Jr. for helping me with his father's papers in Missoula.

The account of Pat Nixon's visit came from old television footage now available on the Web, from news clips in the possession of Jim Brady, from Brady's own photos, and from photos in the collection of Yellowstone National Park.

John Craighead's briefing on his, Frank's, and Joel Varney's population modeling of the Greater Yellowstone grizzlies was documented in an abstract and a text of his speech to the bear meeting on September 19, 1972, at Mammoth Hot Springs, which are available in his papers. They published their completed population analysis through John's unit at the University of Montana in September of 1974, as *A Population Analysis of the Yellowstone Grizzly Bears* (Missoula: Montana Cooperative Wildlife Research Unit, University of Montana, 1974).

Materials on the Second World Conference on National Parks, including the conference proceedings, lists of attendees, schedules and agenda, color slides, and correspondence, are in the collections of the National Park Service at Charles Town, West Virginia. My thanks to John Brucksch for his assistance during my visit there.

29: B-1

I'm grateful to David Graber, Jan van Wagtendonk, Chris Vandiver, the late Galen Rowell, George Durkee, Armand "Herbie" Sansum, Ron Mackie, Doug Robinson, Yvon Chouinard, Rick Smith, Christina Vojta, and Keith Hadley for their help in reconstructing the Yosemite Valley of 1972–76. The account of the fatal accident I describe and the statement of the witnesses about GO CLIMB A ROCK T-shirts are from the report published by the American Alpine Club. Doug Robinson clambered around the cliffs below B-1 with me.

The bear biologist Rachel Mazur has published the definitive work on the "rewilding" of Sierra black bears, including a blow-by-blow account of the long and slow improvement in the designs of bear boxes and backcountry food canisters, in *Speaking of Bears: The Bear Crisis and a Tale of Rewilding from Yosemite, Sequoia, and Other National Parks* (Guilford, CT: FalcolnGuides, 2015). With regard to how intelligent bears are, she quotes my friend the longtime wilderness ranger George Durkee, who says of them: "One must never underestimate an animal that can ride a bicycle."

30: THE DISCIPLE

William H. Drury and Ian C. T. Nisbet's paper "Succession" appeared in the *Journal of the Arnold Arboretum* (54, no. 3 [July 1973]: 331–68).

David Graber's dissertation was published as "Ecology and Management of Black Bears in Yosemite National Park" (University of California, Davis, Institute of Ecology, 1982).

My narrative of the lightning fires following the fiftieth anniversary of Aldo Leopold's Gila Wilderness, in June and July of 1974, was made from the Forest Service's original reports on the fires, Gila National Forest fire atlases, interviews with smokejumpers and firefighters, historical photos, newspaper stories, and field inspections during three visits to the Gila Wilderness. Tom Swetnam's master's thesis, which seems to have been the fulcrum for change, is "Fire History of the Gila Wilderness, New Mexico" (University of Arizona, 1983).

With regard to the findings of the National Academy of Sciences' "Report of the Committee on the Yellowstone Grizzlies" (Washington, DC: National Academy of Sciences, 1974), see Pritchard, *Preserving Yellowstone's Natural Conditions,* and Biel, *Do (Not) Feed the Bears.*

31: THE VERDICT

In August of 1969, before the bear management meeting led by Starker Leopold at Mammoth Hot Springs that September, Glen Cole issued Leopold's Natural Sciences Advisory Committee a briefing memorandum. The memo pointed to an ominous increase in grizzly attacks on humans, from a single one in the 1950s to eleven in the 1960s. Cole pointed to 1959 as the dividing line between these regimes—which, in addition to being the last year of the decade, also happened to be the year the Craigheads began working with bears in Yellowstone. As Frank Craighead testified at the Walker trial, the rangers were happy to let the Craigheads, who were pioneering capture drugs, take bears out of developed areas for them. Cole's memo indicated that in 1959 came the change from killing problem bears to relocating them. He wrote that this was the problem: bad campground bears were allowed to remain in the population because of the new technologies and research methods. Cole recommended terminating bears that were causing problems earlier in the process. He assured the committee that these bears would be replaced by new cubs, which under natural regulation were produced at a rate that would adjust itself in accordance with the population's need.

Aldo Leopold's comment on tinkering is in his 1938 essay "Conservation," in *Round River.*

32: The Appeal

The appeal court's decision is 546 F.2d 1355, *Dennis G. Martin, Administrator of the Estate of Harry Eugene Walker, Plaintiff-Appellee, v. United States of America, Defendant-Appellant.* William Spivak did not write the appeal. It was turned over to the US Attorney's Appellate Division, where it was handled by Robert Glancz and Judith Hale Norris. In addition to those points I have covered briefly in the text, Glancz and Norris argued that the Park Service had no duty to insure imprudent people who would camp illegally and leave their food on the ground. Secondly, the service had no duty to warn Harry Walker about the presence of grizzly bears at Old Faithful, because he had not paid to enter the park and was therefore not an invited guest of the establishment, and because, by walking off the trail at Old Faithful, he had exceeded the geographical limits within which even invited guests were supposed to remain.

Epilogue

Much of Alston Chase's *Playing God in Yellowstone: The Destruction of America's First National Park* (Boston: Atlantic Monthly Press, 1986) was to be in Charles Kay's master's thesis, "Yellowstone's Northern Elk Herd," (ibid.), which is worth reading by itself, if you are comfortable with dry scientific prose. Kay got it right. The Northern Range was crippled.

Toward the end of Chase's *Playing God* (344–62), to illustrate the new druidism that he felt had infected the National Park Service and environmentalism in general, Chase visits California philosophy professor George Sessions, coauthor of a popular book on the "deep ecology" of Norwegian philosopher Arne Naess. To get a sense of what the arch-conservative Chase picked out as the detestably unscientific nature-spiritualism behind natural regulation, see Bill Devall and George Sessions, *Deep Ecology: Living as If Nature Mattered* (Layton, UT: Gibbs Smith, 2001).

By 1986, when Chase's book came out, Yellowstone had been embattled by one controversy after another: the Big Kill, the Craighead era, the listing of the grizzly under the Endangered Species Act, bison, and brucellosis. This continued with Chase's criticism and worsened still after the 1988 fires. Mary Meagher, a remarkable woman who courageously entered a boys' club of wildlife biology when she became Starker Leopold's graduate student in the early sixties, was often the only female PhD in any room full of managers and scientists, whose wives had been told to stay at home. In the sixties at

Yosemite, Ranger Ron Mackie, whose wife had picked up a job with the park hotelier, was pointedly told that it was unseemly for rangers' wives to work and was asked to have his wife resign. She refused.

Meagher succeeded Cole in 1976 as the grizzly population continued to dwindle, and spent her years as chief biologist moving from one political crisis to the next: burgeoning elk, dwindling grizzlies, bison accused of bringing brucellosis to domestic cows, and then fire. She became the tough lieutenant of a Fort Yellowstone under siege. Not surprisingly there was a defensiveness to the Yellowstone scientists' conduct, and not without good reason. The feeling of being embattled makes people stick to their positions. Yellowstone scientists were still defending natural regulation on the Northern Range into the 2000s, when the regrowth of willows, aspens, and cottonwoods, and the appearance of Bill Ripple and Bob Beschta, finally put an end to that line of thinking.

If a new era was coming, it began in Yellowstone's way of dealing with the press's sensationalistic treatment of the 1988 fires as unmitigated disaster. Superintendent Bob Barbee hired Paul Schullery, a fine historian and writer, to produce deeply thought treatments for the press, parsing out the ecological meaning of the fires. Yellowstone issued press releases about the return of life to the park, and welcomed scientists to study the big event, hosting scientific conferences on its physical and biological outcomes. It was a new day. A sense of openness and commitment to dialogue prevailed.

For my very short account of the 1988 fires I am grateful for my conversations with Bob Barbee. I also reference Hal Rothman's *Blazing Heritage,* David Carle's *Burning Questions: America's Fight with Nature's Fire* (Westport, CT: Praeger, 2002), Rocky Barker's *Scorched Earth: How the Fires of Yellowstone Changed America* (Washington, DC: Island Press/Shearwater Books, 2005), Mary Ann Franke's *Yellowstone in the Afterglow: Lessons from the Fires* (Mammoth Hot Springs, WY: Yellowstone Center for Resources, Yellowstone National Park, 2000), and the proceedings of a 2008 conference in Jackson Hole—a twenty-year retrospective—on the fires, *The '88 Fires: Yellowstone and Beyond* (Tallahassee: Tall Timbers Research Station, 2009).

Ripple and Beschta's early work is covered in William J. Ripple and Eric J. Larsen, "Historic Aspen Recruitment, Elk, and Wolves in Northern Yellowstone National Park, USA" (*Biological Conservation,* no. 95 [2000]: 361–70). For their analysis of how wolves might change the feeding behavior of herbivores and thereby affect the growth and distribution of plants and forests, see William J. Ripple and Robert L. Beschta, "Wolves and the Ecology of Fear: Can Predation Risk Structure Ecosystems?" (*BioScience,* 54, no. 8 [2004]: 755–66). A suite of the trophic cascades work of Ripple and Beschta and their associates can be downloaded at the Oregon State University,

College of Forestry, Trophic Cascades Program website: www.cof.orst.edu/cascades. Note especially their work on the effects of herbivory on songbirds and on beaver. See also Charles Kay (ibid.).

Science writer Will Stolzenburg covers Ripple and Beschta's work in a chapter "Valley of Fear," in *Where the Wild Things Were: Life, Death, and Ecological Wreckage in a Land of Vanishing Predators* (ibid., 134–55). The valley in the chapter's title is the Lamar. Other popular treatments of their work are Jim Robbins, "Lessons from the Wolf" (*Scientific American,* June 2004, 76–81) and Jordan Fisher Smith, "Destination Science: Yellowstone National Park, USA" (*Discover* [April 2010]: 50–52).

Notwithstanding its criticism of the word *natural,* the National Research Council study published in 2002, David R. Klein, et al., *Ecological Dynamics on Yellowstone's Northern Range* (Washington, DC, National Academies Press, 2002) is an example of the persistence of the idea that there was nothing wrong with the Northern Range, even as Ripple and Beschta and others were watching the ecosystem disprove that notion after wolf reintroduction.

Among the many books on Yellowstone wolves, I recommend a memoir of wolf reintroduction, one of the two authors of which is Doug Smith, the Park Service scientist who carried it out: Douglas Smith and Gary Ferguson, *Decade of the Wolf: Returning the Wild to Yellowstone* (Guilford, CT: Lyons Press, 2005). My thanks to Doug for all the time he spent explaining things to me during our visits, phone calls, and exchange of e-mail. Notably, he was a coauthor on one of Ripple and Beschta's early papers when many other people in the Park Service were still dismissing the two Oregon researchers.

In my discussion of wolf reintroduction, its effects on the northern elk herd, and of bison and brucellosis I referred to Jerry Johnson, ed. *Knowing Yellowstone: Science in America's First National Park* (Lanham, MD: Taylor Trade Publishing, 2010), and P. J. White, Robert A. Garrott, and Glen E. Plumb, eds., *Yellowstone's Wildlife in Transition* (Cambridge: Harvard University Press, 2013). On the problems of managing a natural population of bison in a limited area under pressure from surrounding agricultural interests, see the chapter by P. J. White, John J. Treanor, and Rick L. Wallen, "Balancing Brucellosis Risk Management and Wildlife Conservation." Wallen is the park's bison biologist, whom I would also like to thank for his patience with my questions.

Lee Whittlesey and Paul Schullery's magnum opus—in which they set out to correct Alston Chase and Charles Kay's allegation that before the Park Service remanufactured it, Yellowstone was a land of scarce game, is Lee H. Whittlesey, Paul D. Schullery, and Sarah E. Bone, *The History of Mammals in the Greater Yellowstone Ecosystem, 1796–1881: A Cross-Disciplinary Analysis of Thousands of Historical Observations* (publication in process).

For a review of human injuries and fatalities in Yellowstone bear attacks, see Kerry A. Gunther, "Risk, Frequency, and Trends in Grizzly Bear Attacks in Yellowstone National Park" (*Yellowstone Science*, 23, no. 2 [December 2015]: 62–64). See also the National Park Service's list of fatal attacks at Yellowstone, www.nps.gov/yell/learn/nature/injuries.htm and at Glacier National Park, www.nps.gov/glac/learn/news/upload/Bear-Deaths.pdf.

INDEX